THE BULL SHARK COMPENDIUM

BY

Zachary Webb Nicholls

Printed By
DEEP SEA PUBLISHING
According to the design of
Z.W.N., 1st DR. JAWS

2017

THE BULL SHARK COMPENDIUM

The Bull Shark Compendium makes transformative and non-substitutive use of multiple scientific (that is, empirical and non-creative) papers for commercial purposes. These aforementioned papers are completely cited, and credit for their original content is exclusively due to their respective authors and publishers. Furthermore, access to the content of these papers has already been made free to the general public; no access to the papers featured within *The Bull Shark Compendium* was acquired by commercial means. While *The Bull Shark Compendium* addresses multiple scientific topics, the work itself is not empirical in nature and does not claim to add to the public body of scientific knowledge; rather, *The Bull Shark Compendium* serves as a creative, original and opinionative address of science, philosophy and aesthetics that makes fair use of the aforementioned scientific papers for narrative purposes. The author and publisher do not claim ownership or authorship of any of the aforementioned scientific papers featured within *The Bull Shark Compendium*.

ISBN-13: 978-1939535894

ISBN: 1939535891

www.deepseapublishing.com

Printed in the United States of America

TABLE OF CONTENTS

To Messrs. Tibbett, Godschalk & Walton
Who were there in the beginning

And to the Brock sisters
Hayley, Amanda, & Lindsey
By whom I am inspired

The Bull Shark pulls flesh
To compound its heaviness
And push the waters

Dear Samuel,

It is 2015, and much has changed, though the blood of your world still pumps its dark intrigue, if not subversively. Empiricism now dominates and sheds an inexplicably brilliant light on former mysteries, revealing much about our beloved designs and their greater purpose in the ecological scheme. Note, however, that I have chosen this word 'scheme' deliberately, Samuel, for it has become apparent that this white, crystal-clear narrative of science is, in fact, a perfect design of an imperfectly designed being: a four-dimensional, five-sensed entity of still-developing mental capacity and (seemingly) decreasing spiritual cultivation. We have shifted greatly, in that our trust is almost purely devoted to all things logical and provable; this is of course a largely positive development, but it has the debilitating drawback of denying the possibilities beyond empirical thought.

You, Samuel, obviously know that I lack faith in all things challenging science: to challenge our objectivity would be idiotic, irrational, and perhaps even immoral. I do, however, accept that our logical minds only *understand* an incomplete part of the grander idea of being. Metaphorically, I believe as if every subject in the universe is composed of two halves of the human conscious: a white and a black (please imagine, for a pleasing aestheticism, a Chinese taijitu). The white is the side of logic and reason—of truth universal and incontrovertible, illuminating all peoples, cultures, and faiths with the crystal beams of knowledge—in other words, science and empiricism. The black, in perfect complement, is the side of faith and feeling—of beliefs NOT absent of empiricism, but rather, BEYOND its limited confines. In the black dwell the arts and philosophies, for it is the realm of the swooning unknown, no less captivating than its white brother. Indeed, in order for something to be truly *understood*, the two must be embraced as one.

In this harmonious manner, I very dearly wish to *understand* something, Samuel. However, I am a bit apprehensive, in that I feel that by fully committing to this wholesome predisposition, I will embark on a strange, stirring, and potentially dangerous journey. I do not know why danger comes to mind, for indeed if I shall fully embrace the light of empiricism, I would certainly avoid the darkness of mistruth, yes? Not to say I will evade darkness completely, for at times I will indeed enter into a different kind of shadowy place…a noble kind…a kind you knew so well. With you, it was reached through a laudation of the horrors, terrors, and supernatural mysteries of the sea. Though her power was *apparent* to every man familiar with her sight, her smell, and her sound, it was only to you truly *understood*, as evident in your rhymes that so successfully captured a small piece of her soul.

As you loved the entire ocean, Samuel, I shall love but a single elemental: a being profound to my being, in that it somehow exerts within myself a gravitational pull to which I must respond with an orbital push—a realignment into a proper, comprehensive sphere of *understanding*. Thus, I shall compose something unprecedentedly congruous: a scientific compendium of all publically available works on my subject, in which each entry shall not only be completely analyzed and succinctly summarized, but also carefully conjectured with new philosophies that seek NOT to counter the illumination of empiricism, but rather, to complement it through an appreciation of the contrasting darkness of all things beyond our immediate perception—in a word, 'supranatural'. In attempting to reconcile these two halves, I hope to create a marvelously balanced meditation on my mortal muse—an approach that fully respects the being as IS, and what it may MEAN for us to approach it altogether.

These are my intentions for the *Bull Shark Compendium*. In fully describing them to you, Samuel, I feel ready—and perhaps, if it's possible, even *more* invigorated than before—to begin this truly comprehensive exploration of what we have come to call '*Carcharhinus leucas*'. Though I hope to be as fruitful in this endeavor as you were—to reach a parallel depth of *understanding* in regards to this marvelous elemental (and its possible implications)—I will, regardless, rejoice in every instance appropriated by this meditation. It is, after all, the journey that provides the key for the destination's mirrored lock.

Wishing you good tidings, and in a most jovial manner, your student,

The 1st Dr. Jaws

Zachary Webb Nicholls

A Note to You,

You are about to fall through time, and at a velocity near parallel to that of my own descent. I understand exactly what it is that I am seeking, but am completely unaware as to what form it may take shape; as such, I am only an inch or two ahead of you, my friend, in this rapid submersion into the oceans of thought, philosophy, science, and *understanding*—in other words, you and I are roughly 'in this together'. As such, it is imperative that you know the basic attributes of our endeavor, namely:

1. That we shall chronologically venture from discovery to discovery as they relate to our subject and muse: the elemental we know simply as *Carcharhinus leucas*.

2. That each point of contemplation shall be represented by a single pair of pages: a Page of Science and a Page of Conjecture

3. That the Page of Science, on the left, shall be composed of a published abstract or introduction of an original scientific work reporting on a relevant, publically available discovery as it relates to *Carcharhinus leucas*. The date and original authorship shall be noted, and the text will remain completely unaltered (save for the purposes of abbreviation), even in the cases of a non-English language.

4. That the Page of Conjecture, on the right, shall be composed by myself, and will serve as a reflection on the content of the page opposite. This meditation at times may be philosophical, prosaic, or even picayune, but nonetheless will serve as a guide to you, my friend, in cases both synonymous and antonymous to your own ideals.

On this latter point, always remember to form your own conjectures as it relates to *Carcharhinus leucas* and any of the new, exciting, and potentially unsettling ideas that we may come to face on this adventure. Science is truth within our perception, so never doubt it. Conjecture is idea beyond our perception, so never ignore it. Trusting in one alone is foolish, but embracing them together is wise. Be open-minded as we explore, and embrace the discoveries, challenges, and ideals that may relate to our muse, our grander idea of being, or even yourself.

Thank you for accompanying me on this venture. I use the word 'venture' justly, as I'm not entirely sure how it will end. Indeed, as you are reading this now, I have not yet written a single page on that which lies ahead; it is now January 27th, 2015, and this date marks the very beginning of what will become...

THE BULL SHARK COMPENDIUM

1841

Systematische Beschreibung der Plagiostomen
Berlin, Veit, pp. 1-200
MÜLLER, J. & HENLE, F.G.J.

Spec. 17. Carcharias (Prionodon) leucas. Val.

Kopf.   Schnautze sehr kurz und stumpf. Naslöcher sehr nahe am vorden Ende der Schnautze. Naslöcher und Augen gleich gross.

Zähne.   Zähne oben dreieckig, die äussere Seite mit einer geringen Spur von äusserem Winkel, untere Zähne gerade, schmale Pyramiden auf breiterer Basis, Schneide gezähnelt. Zahl der Zähne 27/27.

Kiemen.   Zwei Kiemenlöcher über der Brustflosse.

Flossen.   Erste Rückenflosse dicht hinter der Basis der Brustflossen, mit vorderem und hinterem spitzen Winkel, ausgeschnitten. Das hintere Ende der Afterflosse geht beträchtlich über das Ende der zweiten Rückenflosse, so dass das Ende der letztern auf das letzte Drittheil oder gar die Mitte der Afterflosse fällt. Brustflossen sichelförmig, mit äusserm spitzen, innerm spitz abgerundeten Winkel.

Farbe.   Farbe weiss, oben und unten. Die Brustflossen waren an den äussern Winkeln, bei zwei Exemplaren, dunkler ins Bräunliche. Bei allen war der untere Lappen der Schwanzflosse bräunlich und bei zweien der untere Rand der ganzen Schwanzflosse eben so.

Schuppen.   Schuppen drei – bis fünfkielig.

Vier Exemplare trocken in Paris, von den Antillen, durch Plée.

This page on the Bull Shark (here described as *Carcharias leucas*, or "White Shark") is from Müller & Henle's *Systematische Beschreibung der Plagiostomen*: a marvelous German catalogue of 97 sharks and 114 rays arranged in respect to their mid-19th Century Linnaean classifications. Wonderfully succinct, this account features physical profiles in which the following characters are described in crisp detail: Head, Teeth, Gills, Fins, Color, Scales, and Size. An exquisite color plate also accompanies some species (the plates being located towards the end of the *Systematische Beschreibung*), but to our misfortune *Carcharias leucas* lacks such a plate.

Though we do not have a pictorial representation, we still clearly see the foundations of the physical motif we have created for *Carcharhinus leucas*. Key features as originally delineated by Müller & Henle (who are the species authority, officially naming the beast in 1839) include: a very short and blunt snout, triangular upper teeth complementing broad-based pyramidal lower teeth, crescent-shaped (falcate) pectoral fins, and predominantly white color. The color section, in particular, fascinates and confuses me; if I am reading the entry correctly, Müller & Henle remark that the shark is white above and below, and this is especially interesting in regards to the final species name of "*leucas*" (which is Latin for "white", "white-sided", or "white-shielded"). As the shark is grey above and white below in life, I wonder if the four preserved specimens Müller & Henle acquired from Plée's West Indian sample (stored in Paris) where completely white postmortem.

A delightful attribute of the *Systematische Beschreibung* is the large amount of international cooperation required for its success; Müller & Henle gratefully collaborated with scientists from London, Paris, Trieste, and others to form this impressive digest.

1851

List of the Specimens of Fish in the Collection of the British Museum Part I: Chondropterygii

British Museum (Natural History), London, 160 pp.

GRAY, J.E.

THE principal object of the present Catalogue has been to give a complete list of the specimens of Fish contained in the collection of the British Museum, indicating the peculiarities of each, as regards variation of character, locality, and the source from whence it has been derived. `

JOHN EDWARD GRAY

LIST OF FISH.

10.   SQUALUS (CARCHARINUS) LEUCAS.   Leucas Shark.
Carcharias (Prionodon) leucas, *Valenc.*   *Müll. & Henle, Plag.* 42.
HAB.—West Indies.   Mus. Paris.

***   *The first back-fin in the middle between the pectorals and ventrals, or nearer the former.*

ADDITIONS AND CORRECTIONS.

Page 46,

10.   SQUALUS LEUCAS, *add*

a.   Stuffed.   Presented by the Lords of the Admiralty.   From the Antarctic Expedition.

The most remarkable feature of this excerpt from Gray's catalog is the inclusion of four separate genus names for our shark (here known as the "Leucas Shark"). *Carcharhinus leucas* (literally translated as the "White Sharpnose") is the currently valid taxon, but it may behoove us to appreciate the outdated synonyms, at least in terms of their English translations and current use:

Carcharias	"Shark"	Currently used for a sandtiger shark
Prionodon	"Front-Tooth"	Currently used for linsang
Squalus	"Sea Fish"	Currently used for dogfish sharks
Carcharhinus	"Sharpnose"	Currently used for requiem sharks

Gray, in attempting to document the key traits for each of the Museum's elasmobranchs, noted that, "The characters of the genera of Sharks and Rays, with the Synonyms, have principally been derived from the work of Professors Muller and Henle." This quote is particularly fascinating as Müller & Henle have likewise acknowledged Gray as an essential reference for their *Systematische Beschreibung*; as such, we are witnessing a piece of a 170-year-old international discourse on our species.

It's easy to romanticize this dialogue—to imagine finely dressed European gentlemen crossing the English Channel, hoping to gain new insight on the fascinating monsters of the deep. I hope that their discussions were friendly, and I wonder if one party ever offered the other customary pleasantries—a free meal, a drink, or a tour of their respective city, for example. Then again, we cannot possibly know, and such may assume with equal certainty a professional animosity, distrust, or prejudice between these great British and German minds. Such sociological mystery is, at least to me, compelling.

1858

Memorias Sobra la Historia Natural de la Isla de Cuba,
Acompañadas de Sumarios Latinos y Extractos en
Francés. Tomo 2. La Habana.
Vluda de Barcina, Havana, 2: 1-442, pls 1-19
POEY, F.

198.   *Squalus platyodon*   *Poey*
Vulg. *Tiburon*

C'est un mâle du sous-genre *Prionodon*, long de 2500 mill. –
L'ampleur du corps est plus que médiocre, le cou fourni, le museau très
court et très obtus, l'œil très petit, la pupille ronde, ce qui est dû peut-être à
un accident. La première dorsale, peu recourbée en avant, est presque droite
en arrière, son angle supérieur un peu aigu; son lobe postérieur très court: sa
base antérieure répond au commencement du lobe de la pectorale. Voyez pl.
19, fig. 5, 6. $\frac{12+1+1+12}{12+1+12}$. – 627.

199.   *Squalus obtusus*   *Poey*
Vulg. *Tiburon*

C'est un mâle de 2300 millimètres, décrit d'après un individu
bourré avec intelligence par mon zélé collègue D. Juan Antonio Fabre. Il
ressemble si fort à l'espèce du même sous-genre qui précède, que je me
bornerai à en faire ressortir les différences. –Le corps est plus allongé, car
la base de la première dorsale est séparée de la seconde par deux fois et
demie la longueur de cette même base; la pupille est fendue verticalement;
les deux angles de la pectorale sont aigus; celui de la dorsale antérieure est
plus pointu; la seconde dorsale est plus échancrée. –Voyez pour les dents
pl. 19, f. 7,8. —575.

These two separate accounts of *Squalus platyodon* and *Squalus obtusus* are actually describing the same species; though Poey of course did not know this at the time, his suspicions are evident in the *S. obtusus* entry, as he writes, "It is so like the species of the same subgenus above, I shall confine myself to highlight the differences" (rough translation). The meanings and current uses for these species names are as follows:

1	*platyodon*	"broad-tooth"	Some use, notably Ichthyosaurs
2	*Obtusus*	"obtuse"	Many uses for many different kinds of taxa

I believe these names are in reference to the dentition of each animal, though both indeed exhibit other characteristically broad features (namely the snout and generally stocky body). As they have been later determined to be the same species, there really is very little distinguishing the two: *S. obtusus* was described to have bottom teeth with a broader pyramidal base, dermal denticles with five ridges, and a slightly longer separation between the bases of the first and second dorsal fins, whilst *S. platyodon* had wider upper teeth, dermal denticles with three ridges, and a slightly shorter separation of the dorsal fin bases.

Though it is a bit confusing that we now have six different scientific synonyms for the Bull Shark, there is much value in embracing them as symbols of the entity's core perceivable identity; mistaken or not, the traits that we as humanity have attributed to this being are (and forever will be) "white", "shark", "sea fish", "sharpnose", "front-tooth", "broad-tooth", and "blunt". This is a very specific combination that no other creature shares; there could be "black" "sharp" "bear", or, "land bird" "no-tooth" "owl", but more likely than not, there couldn't be another combination quite like ours. This is uniqueness.

1859

Notes on Sharks, More Particularly on Two Enormous Specimens of Carcharias leucas, captured in Port Jackson, Sydney, New South Wales.
Proceedings of the ZSL, 1859: 223-226
BENNETT, G.

An enormous Shark (*Carcharias leucas*, Valenciennes) was lately captured in Port Jackson by two boatmen, T. Mulhall and J. Rica, who finding him ranging about the harbor, procured a harpoon and went in chase of him. They succeeded in harpooning the monster, who when struck ran away with a great length of line. Being tired, and finding himself fast, he rushed back again and attacked the boat, leaving five teeth broken in the wood. The boat fortunately was strong enough to bear the shock. He then ran off again to some distance, and, finding escape hopeless, rushed a second time at the boat. On this the men attacked and finally succeeded in disabling him by violent and repeated blows upon the head with a large piece of wood; they then towed him the whole length of the line, so as "to drown him," as it is termed, and brought him to Sydney alive, but helpless. He died some hours after being landed on the wharf, being very tenacious of life. The huge monster was soon a great object of curiosity, and, being enclosed, was duly advertised for exhibition to the public whereby the capturers realized the very handsome sum of about £80. The animal was afterwards presented to the Museum, in which institution it remains in an excellent state of preservation. Its size, by actual measurement, is as follows:

The circumference of the body, about the centre [6 ft. 7 in.]

Length from the end of the tail to the point of the nose [12 ft. 4 in.]

This is undoubtedly the most elaborate account of our shark yet, with attention paid not only to physical attributes and specific dimensions, but also to stomach contents, parasites, means of capture, economic use, and biological mutualism with pilot fish (genus *Naucrates*). A phenomenal mid-19th Century narrative to be sure, Bennett's is nonetheless compromised (or, perhaps dear reader, complemented) by the infusion of prosaic stigma: he refers to the shark as "monster" twice, and notes upon his examination of the stomach contents, "I would not venture to send one alive to the Zoological Gardens, as its keep would be ruinous". To his point, however, this particular individual's stomach contained 300lbs of horsemeat, the back half of a pig, the front half of a bulldog, and a ship scraper.

Though there is a modern push against shark stigma, I personally find it enlightening as a contrast to today's more positive narrative of 'shark respect'. Like the prime dualities of light and darkness, stigma and respect are powerful opposites: their relationship, though contrasting on the surface, is one of profound harmonization, in which each component of the duality is the other's greatest complement. In terms of "The Shark", this concept has two significances:

1. Stigma more brilliantly reveals the radiance of respect, and as such is an important perspective to embrace throughout our tour of shark history, especially in light of our modern push for a more positive viewpoint.

2. "The Shark" is a dualistic identity that simultaneously impresses and disturbs the beholder. This duality constitutes a powerful complex that not many other creatures share.

The age of Bennett's account (156 years) may attest to the innateness of this dynamic.

1877

Synopsis of the Fishes of Lake Nicaragua
Proceedings of the Academy of Natural Sciences of
Philadelphia, 29: 175-191
GILL, T. & BRANSFORD, J.F.

GALIOSHINIDÆ

Eulamia nicaraguensis.

The shark of Lake Nicaragua cannot be identified with any of the previously-deseribed forms, although closely related to *Eulamia Milberti* and the kindred species. The speeimen deseribed, when freshly eaught, measured 6 feet 4 inches in length. The skin and jaws were preserved. Althongh, as indieated in the introductory remarks, it has long been known that a shark was an inhabitant of the lake, the relations of the speeies have been previously unknown, and the spoils obtained by Dr. Bransford are the first that have been subjeeted to seientific examination.

Larger specimens than that obtained might have been proeured, but Dr. Bransford took the first eaught. There are numerous well-authentieated eases of people having been killed by these sharks, and the natives are very careful to keep out of their way. Some six months before Dr. Bransford's arrival a man was bitten by one near the place where this speeimen was taken. A great portion of one thigh and buttock was eut away, and he died from the effeets. Repeated tales are told of similar ineidents. Squier says: "Sharks abound in the lake. They are called tigrones from their rapaeity. Instances are known of their having attaeked and killed bathers within a stone's throw of the beach at Granada."[1] They are found throughout the length of the river San Juan, of sizes varying from a foot to over six feet in length. Sivers thinks that they eome up the river from the sea.

The Lake Nicaragua Shark: its very name summons exotic and nightmarish impossibilities, i.e., that a shark—a terror once considered exclusive to the ocean depths—could haunt a freshwater lake in abundance. What's especially chilling about this concept is the fact that *Eulamia nicaraguensis* ("True Shark from Nicaragua") is not an innocuous species, but instead *our* shark. This record is 138 years old—half a century older than the International Shark Attack File—and yet is already a sound testament to the Bull Shark's lethality. As opposed to being a trivial subject of freak accidents, this species is an assuredly threatening presence, bearing a legacy of multiple attacks so thorough that, "the natives are very careful to keep out of their way".

Please understand that I am not trying to excitedly "play-up" the dangerousness of the Bull Shark so as to satisfy a perverted need for pornographic violence (a need well-exploited by most modern shark television programs). Instead, I want to examine the significance of such a violent reputation and its true importance: that this shark is *impactful*. It's a cultural force—an argument for a people to alter their way of life, namely, "keep out of their way"—and this grants the species a certain sense of power in our human existence. The Bull Shark in its very nature demands from us a specific address; we as a species must acknowledge that this entity is a death-bringer, and that furthermore it can hunt mostly undetected within its aquatic medium (thus creating a "false ubiquitousness"—meaning that as we cannot triangulate its exact position with certainty, it can BE anywhere, and thus FEEL everywhere).

Of course, the Bull Shark is not and cannot be limited to only these alarming personas, but they do play a very significant role in our muse's characterization. It's a complex and dark dynamic that is (maybe not so strangely) alluringly beautiful.

1879

Manuale Ittiologico del Mediterraneo ossia Sinossi
Metodica delle Varie Specie di Pesci Riscontrate sin qui
nel Mediterraneo ossia Senossi Metodica delle di Sicilia.

5 Parts.

Palermo: Pt 1, 1879:i-viii + 1-67; pt 2, 1881:1, -120; pt 3,

1884:121-258

DÖDERLEIN, P.

18.	Carcharias *(Prion.)*, **Leucos,** Val. (mss.).	
Sp.	Müller Henl., Plagiost., p. 42.	1851 *Squalus*
1841	Guichenot, Poiss. d'Algerie, p. 124.	*(Carcharinus)*
1850	Bennett, Proceed. Zool. Soc. London, p.	*Leucos,* Gray, List.
1859	223.	Chondropt. Brit.
1865	Dumeril, Elasmobr., p. 358, sp. 6, et p. 143.	Mus., p. 46, n. 10.

Italiani		
Carcaria	Mari delle *Antille.*	*Denti della masc. sup.*
bianco (Ital.).	*Mediterraneo* accidentale.	Triangolari dentellati,
Esteri	Un individuo venne	inclinati all'esterno con
Requin blanc	ritrovato da Guichenot	picc. Angolo rientrantenel
(Fr.).	presso le coste d'Algeri.	margine esterno.
Leucos Shark	È una delle specie più	
(Ingl.).	voraci, che attingè alla	*Denti della masc. infer.*
	lunghezza di 4^m e più.	stretti, acuti, a base larga,
		coi margini dentellati.

Perhaps the most intriguing feature of the *Manuale Ittiologico* is Döderlein's self-awareness and humility: traits of a truly enlightened individual. Specifically, Döderlein introduces the *Manuale Ittiologico* with the assertion that it's an incomplete work; some accounts may only be based on little or false information (i.e., a misguided affirmation by local fishermen or other laypersons), while other scientifically credible entries are still rendered to be non-definitive. This is actually a mark of genius, as Döderlein's mission was to simply compile all existing information on Mediterranean Fishes (regardless of source) so as to provide a complete compendium with which the scientific community can easily reference, discredit, or amend its entries. In doing so, Döderlein effectively eliminated the need to waste time scouring through defunct or faulty information from broad sources, thereby facilitating international scientific progress towards a more complete understanding of the Mediterranean's fish species. In short, each entry is merely a launch point to inspire further discussion and study.

The Bull Shark account (listed under the slightly different scientific name of *Carcharias (Prionodon) Leucos*) is rather small compared to those of the three other *Carcharias* species described. It yields to us few surprising details (namely, the confirmation of a slight Mediterranean presence, as well as the contemporaneous Italian, French, and English common names). As you can see, Döderlein references our now-familiar sources of Müller & Henle, Gray, and Bennett for this entry; from those authors, he confirms only the species' important physical attributes and voracious appetite. I have chosen to just highlight upon the descriptions of teeth, so as to both save room and highlight their importance in distinguishing a shark species (again, Bull Sharks have triangular uppers and narrower, broad-based pyramidal lowers; both sets are serrated).

1882

Notes on Fishes Observed about Pensacola, Florida, and Galverston, Texas, with Description of New Species
Proceedings of the United States National Museum,
5: 241-307
JORDAN, D.S. & GILBERT, C.H.

3. Carcharias platyodon (Poey.) J. & G. *Shovel-nosed Shark (Galveston).*

This is the commonest of the large sharks found on the coast of Texas in the summer. A young male specimen 32 inches long was obtained at Galveston, and the jaws of a very large example, in the possession of Mr. E. Gabriel, of Galveston, were also examined.

The specimen here described was not preserved, it having spoiled before the arrival of alcohol.

Among the described species of this genus Carcharinús platyodon *(Poey)* (*=obtusus* Poey) seems to be the most nearly related to the species examined by us. The pectoral in *C. platyodon* is larger, the teeth somewhat different, and the second dorsal is said to be "assez grande," whereas in *C. cœruleus* the latter fin is very small. *C. fronto*, lately described by us from Mazatlan, is also very similar, but has a much larger second dorsal.

Another species, similar, but with longer snout, has been described by Dekay under the name of *Carcharias cœruleus*. This description has been referred by Professor Gill to the synonymy of the very different species, *Carcharias plumbeus (Nardo)* = *Carcharias milberti* M. & H., and has been called "*Eulamia milberti*".

There is, however, no good evidence that *C. milberti* (*plumbeus*) has ever been taken in our waters.

Jordan & Gilbert list seven species of *Carcharias* as residents of the American Atlantic and Gulf coasts. They are (in conjunction with their current names):

OLD NAME	NEW NAME	COMMON NAME
Carcharias glaucus	*Prionace glauca*	Blue Shark
Carcharias obscurus	*Carcharhinus obscurus*	Dusky Shark
Carcharias cœruleus	*Carcharhinus plumbeus*	Sandbar Shark
Carcharias platyodon	***Carcharhinus leucas***	**Bull Shark**
Carcharias limbatus	*Carcharhinus limbatus*	Blacktip Shark
Carcharias brevirostris	*Negaprion brevirostris*	Lemon Shark
Carcharias terræ-novæ	*Rhizoprionodon terraenovae*	Atlantic Sharpnose

All aforementioned species are now classified as members of the family Carcharhinidae (Requiem Sharks), and the list of described Carcharhinids from the American Atlantic and Gulf Coasts has more than doubled since Jordan & Gilbert's original tally. The species depicted here as *Carcharias platyodon* is very likely a Bull Shark, but there is an understandable air of uncertainty. After all, *Carcharias platyodon* has been historically misapplied to the Sandbar Shark (*Carcharhinus plumbeus*), and furthermore this individual of Jordan and Gilbert's could only be compared to seven (out of about 15 now-known) American Carcharhinid archetypes. This, in addition to the unending Linnaean "name game" of shifting taxa, indeed makes for a confusing challenge to proper species identification.

The only other thing I'd like to note here is the attention to this young individual's "distinctly bluish tinge above" and "numerous mucous pores" (Ampullae of Lorenzini).

1883

Faune de la Sénégambie. Poissons
Paris :O. Doin, 1883-1887
DE ROCHEBRUNE

CONSIDÉRATIONS GÉNÉRALES.

§ I. – Depuis l'époque où A. Duméril écrivait son important Mémoire sur les Reptiles et les Poissons de la côte occidentale d'Afrique (1), peu de travaux sont venus accroître le nombre des espèces décrites ou mentionnées par le savant Professeur du Muséum de Paris.

Parmi les rares publications relatives à l'Ichthyologie, nous trouvons seulement les brochures de M. Steindachner sur quelques espèces du Sénégal (2), et les notes de MM. Péter (3), Gill (4), Cope (5), Gunther (6), relatives au meme sujet.

Nous indiquerons, outre ces divers ouvrages, la Faune des Canaries, par Valenciennes (1), le Traité de pêche sur la côte occidentale d'Afrique, par Berthelot (2), le mémoire de Castelneau (3), sur les Poissons de l'Afrique austral, les ouvrages d'Adanson, don't les deux principaux (4), renferment de précieux renseignements; enfin les traits généraux de Lacépède (5), Cuvier et Valenciennes (6) et quelques notes de Guichenot (7).

8. CARCHARIAS (PRIONODON) LEUCOS. Dum.

Carcharias (Prionodon) leucos, Dum. Elasm., p. 358.

Dekojh. – Assez commun : --Dakar, cap Vert, baie de Sainte-Marie, marigot de Saloum.

I was at first surprised to see the abbreviation "Dum." (for the French zoologist, André Marie Constant Dumeril) next to the species binomial, as this would indicate that Dumeril was (incorrectly) the species authority. However, upon referencing back to Döderlein's profile of *Carcharias (Prionodon) Leucos* in the *Manuale Ittiologico* (page 12), I wonder if it was perhaps orthodox in the late 19th Century to consider Dumeril as such (along with Guichenot, Bennett, and the true authorities: Müller & Henle). Conversely, De Rochebrune could be using Dumeril as merely a reference, or—much to the chagrin of modern scientists—as the true, single species authority of our shark, which is very much incorrect (even for that time, I would imagine, as Müller and Henle named the species 44 years before this account).

In any case, *Faune de la Sénégambie* hardly offers us any new information on our shark. Only local haunts are reported (Dakar, Cape Verde, Bay of St. Mary, backwaters of Saloum) along with the fact that the species is "fairly common" in these waters. Saloum is a Senegalese river, which is particularly fascinating as this is our first confirmation of the shark utilizing West African estuarine tributaries. *Dekojh* appears to be the shark's regional name, but of this I am not certain.

Though lacking in Bull Shark information, *Faune de la Sénégambie* is nevertheless an incomparably gorgeous record of West African fauna, with richly illustrated plates of over 100 mammals, birds, reptiles, amphibians, and fish (including 2 rays). Some shark species such as the Scalloped Hammerhead (here listed as *Zygæna leeuwenii*) are presented with a comparably more extensive account. However, most sharks in the *Faune de la Sénégambie* are—like our Bull—treated with brevity.

1886

A Preliminary List of the Fishes of the West Indies
Proceedings of the US National Museum, 9: 554-608
JORDAN, D.S.

In this list I have endeavored to represent the present condition of our knowledge of the fish-fauna of the West Indies. I have included in it all species which have been accredited by good authority to the waters of the West Indies proper and the Bermudas, as well as to the Atlantic coasts of Mexico, Central America, Venezuela, and Guiana. I have excluded from it all species which have not yet been found farther south than the Florida Keys and the "Snapper Banks" of the Gulf, as well as all those as yet known only from Brazil, although, as a matter of course, many of each of these categories will be found to be genuine members of the West Indian fauna, the "Snapper Bank" fauna especially being entirely West Indian in its general character. In most of the families an attempt has been made to exclude purely nominal species, but in some groups (*Siluridæ*, *Syngnathidæ*, *Murænidæ*, etc.), in which no critical studies have yet been made, this has been impossible.

It should be clearly understood that this is simply a preliminary list, which must needs be greatly modified when the species of the different groups receive thorough study. It is probable that comparatively few of the larger shore-fishes are to be added to the list, but of the smaller fishes, and especially of those found in deep water, it is not unlikely that the majority are still undescribed.

14.   **Carcharhinus leucos** (Valenciennes).

(*Carcharias leucos* Val. In Müller & Henle, Plag. 42.)

Alas, another list. I believe, dear reader, that it might be time to take a break from our shark and focus on an equally dark and fascinating subject: David Starr Jordan, the author of this *Preliminary List*. Now I am by no means an authority on the matter, but as modern legend has it, Jordan was involved with a potential cover-up for murder.

110 years ago, Jordan was the president of the newly founded Stanford University (established in 1891 by Leland and Jane Stanford). After Leland Stanford's death, Jane essentially assumed total control of the University's operations. Though Jordan—a nationally respected educator, ichthyologist, and (disturbingly) eugenicist—was himself instrumental in faculty recruitment and program establishment, it was Jane who had the final say in University matters. As such, when tensions between the two bitterly increased after years of discordant cooperation, Jane had the mind (and the power) to ultimately remove Jordan, and was seriously considering doing so by 1905.

It was in that year that Jane had suffered an attempt of murder by strychnine poisoning. Though she survived this first attack, she was deeply troubled at the idea of an unknown attacker, and subsequently retreated to Hawaii to regain her health. However, in Honolulu Jane became the victim of a second attempt via strychnine poisoning, and this second strike had unfortunately claimed her life.

The cause of death was quick for professional doctors to deduce, and was almost publically accepted...save for Jordan. Upon hearing the news, Jordan rushed to Honolulu, found a local doctor, and suspiciously insisted that a cause of death other than poisoning was more likely the case. Upon that doctor's compliant ruling that Jane Stanford died from a heart attack, Jordon quickly publicized this result, which became the final verdict until almost a century later, when old suspicions gained new light.

1896

The Fishes of North and Middle America: a Descriptive
Catalogue of the Species of Fish-Like Vertebrates Found in
the Waters of NA, North of the Isthmus of Panama. Part I.
Bulletin of the United States National Museum, 47: 1-1240
JORDAN, D.S. & EVERMANN, B.W.

48. CARCHARHINUS LAMIA,* (Rafinesque).
(CUB-SHARK; REQUIN; REQUIEM; LAMIA.)

Head broad, depressed: snout short and rounded, nostrils midway between its tip and the front of the mouth; breadth of the mouth $2\frac{1}{3}$ times length of pectoral part of snout. First dorsal very large, inserted close behind the base of the pectoral, its height a little greater than the length of its base, its anterior margin convex, its upper angle rounded, its posterior border nearly straight, its lower angle pointed, its height about equal to greatest depth of body; second dorsal much smaller than first, about equal to anal; pectorals at least twice as long as broad, 5 times in body; upper lobe of caudal $\frac{1}{4}$ the total length, twice the inferior lobe. Grayish, fins rarely darker at tip. L. 10 feet. Tropical parts of the Atlantic; common northward to Florida Keys, abundant in the Caribbean Sea and in the Mediterranean; a man eating shark, notorious in warm regions as a greedy scavenger about wharves. (λαμια, lamia, sea-monster, from λαμος, devouring hunger.) - EU

Carcharhinus leucos, (Valenciennes): Pectorals rather long, but shorter than in *C. lamia* ; first dorsal with pointed angles, its anterior border not convex, and its posterior border little excavated (Duméril); otherwise about as in *C. lamia*, with which it is probably identical. West Indies; Algiers. (λευχος, white.)

20

Jordan's short behavioral notes are fascinating. After reading Bennett's account of the 'Australian monster' and Gill & Bransford's record of the 'Nicaraguan man-eaters', I feel as if Jordan's Cub Shark profile somehow cements the species' notoriety. The Bull is once again confirmed to be a "man eating shark", and further personified as a "greedy scavenger". This latter statement suggests an unrestrained selfishness about the animal—as if its ignominious hunger could not be placated by mere satisfaction with the living, but only through a perverse appropriation of the dead as well. The word 'scavenge' is also a prosaic vehicle of abhorrence, as it summons to mind an ultimate idea of desperation commonly embodied by the repulsiveness of vultures, the ravenousness of wolves, and 'obscenity' of scroungers.

Whether or not Jordan's use of these terms straddles the prosaic and scientific realms is up to interpretation (especially since scavenging is, of course, a sound means of natural nourishment and widely adopted in nature). However, I find it compelling to consider the literary positions of such terms, as doing so discloses a deeper cultural dimension to our shark. As a documented hazard, the Bull Shark has (to nobody's surprise) attracted a negative reputation; however, the severity of this negativity—with "greed", for example, being such an unnatural imbalance of desire that it is often deemed as sinful or evil— is a bit unsettling. It's as if the animal has transcended beyond its natural role of 'predator' into a supernatural place within our psyche—a position of true spiritual consequence where the shark represents a transgression of order.

This idea of transgression wrongly attributes the Bull Shark as sinister or evil in its *self*; the gravity of such attribution is, however, intriguing. How can something *other* than our selves be so easily misrelated to the darkest corners of our own humanity?

1904

The Fishes of Panama Bay

Memoirs of the California Academy of Sciences, 4: 304 p.

GILBERT, C.H. & STARKS, E.C.

8.   **Carcharias azureus**   sp. nov.   CAZON AZUL.

This species is well known though not abundant at Panama, and is more highly prized as food than other sharks. It appeared in the market on two occasions during the stay of the expedition, and three specimens were preserved, measuring from 92 to 95 cm. Two of these are males with the claspers quite undeveloped, not nearly reaching margin of ventrals. The species is said to reach a large size.

C. azureus is extremely near *C. nicaraguensis*, from Lake Nicaragua and its outlet, the San Juan River. Dr. Jordan has kindly compared the above description with a specimen of *C. nicaraguensis* (No. 39913) in the United States National Museum. The latter has a longer and wider snout, the length of which is contained $1\frac{1}{2}$ in its pectoral portion, its width opposite the nostrils equaling the distance from the angle of the mouth to the third gill-slit. The base of the first dorsal is $\frac{1}{2}$ the interspace between dorsals, and the base of the second dorsal is contained $2\frac{1}{3}$ times in the first. The lower caudal lobe is contained $2\frac{2}{5}$ in the upper lobe. The pectoral is but faintly dusky. These differences are not great, but there has been no opportunity to make a direct comparison. In view of the exceptional distribution of *C. nicaraguensis*, known only from fresh waters, which belong to the Atlantic slope, it has not been thought wise to make the identification.

Consider heterotrophy: the consumption of others. We—like sharks and every other animal—cannot produce food for ourselves, and so must take it from the bodies of others in order to survive. It's a brutal, paradoxical system: life is forced to end life in order to continue life.

In this incredibly complex, poetic, and terrifying dynamic, we come to some very curious circumstances; for example, anything we humans eat becomes (at least in part) incorporated into our own body structure. Material that once belonged to a cow, a fruit, a grain, or (as this article suggests) a shark can be transformed into a human hair, muscle, or bone through the powers of digestion. With this in mind, consider the purchase of one of these more valuable Panamanian Bull Sharks from the fish market: flesh belonging to our shark could not only belong to you, but also *become* you. The molecules that once composed a Bull Shark could very well reside in some muscle, tissue, or fiber of your own being. Likewise, anything that a Bull Shark eats—including a human victim— could reside in its own muscles, tissues, or fibers.

This is interplay of form. It's not unique to the human-shark dynamic, but it still plays an interesting role in the relationship; a shark's molecules could become human and a human's molecules could become shark. This exchangeability is profound in multiple ways. Most obviously, the idea of relatedness—that two biological machines could be so similar that they can 'trade parts' and still function without issue—becomes resoundingly clear. Less apparent is the poetry and philosophy of consumption: physical forms are constantly changing and incorporating pieces of *other* physical forms in order to protect the structural integrity of their '*selves*'. What, then, is a selfish *self*? Can it be a spirit requesting parts for its vehicle? Can it be a complex product of the complex of pieces?

1905

Notes on Bermudian Fishes
Bulletin of the Museum of Comparative Zoology at Harvard College, 46 (11): 108-134
BARBOUR, T.

The material which forms the basis for this paper belongs to the Museum of Comparative Zoölogy and is from several sources: first, a collection made by my brother, Mr. W. W. Barbour, and myself during parts of March and April 1903; secondly, a large and rather complete collection made during part of June, July, and part of August, 1903, while I was attached to the Biological Station at Flatts, Bermuda; thirdly, a number of specimens in the collections of the Museum of Comparative Zoölogy, and finally, a series obtained by Professor Mark and collected at the Station during the summer of 1904. I here express my gratitude to Dr. Mark for his kindness in procuring a number of most interesting specimens, and thank Messrs. H. B. Bigelow, Owen Bryant, and J. T. Nichols, for their kind aid in collecting and preserving the largest collection. Finally, it is a pleasure to thank Mr. Samuel Garman of the Museum for the assistance he has giving me in making the identifications.

Carcharhinus platyodon (Poey).   Puppy Shark.
Distribution. — Coasts of Texas and Cuba.
Very common off the Challenger Banks and outside the reefs. Considered a fine food fish by the colored people. The only specimen preserved was identified by Mr. Garman as belonging to this species.

It's very odd to see the term 'colored people' in a professional scientific publication, especially in consideration of the fact that its use was normal for the United States at the turn of the twentieth century. Though we now identify the phrase as a piece of archaic racism, its application 110 years ago was usually inoffensive. This publication's intentions for the term appear to be innocent enough, but the term 'colored people' is in itself a divisive and insultingly simple label for a very complex demographic. To commonly group a people based only on one phenotype—especially a phenotype that is in no way exclusive to a single ethnicity or culture, but instead shared by an incredibly diverse array of cultures and backgrounds from across the globe—is simply unintelligent. The fact that this and other professional scientific publications of the day were able to safely incorporate such simple-mindedness in their work is rather unsettling, and warrants thanks for today's vastly improved sense of social intellect.

The rest of this account offers little new insight; the shark's palatability (previously remarked upon by Gilbert & Starks in *The Fishes of Panama* Bay) is further confirmed, and a preference for seaward reef habitat in Bermudan waters is also established. This latter note on the Bull Shark's haunting grounds (pun slightly intended) summons a powerful image to mind: a ghostly white plain of sand and shells with no other feature than dark, prowling silhouettes fading into and out of the ultramarine backdrop. On sunny days, the light would ripple across onyx backs, only to later roll off into the curling sands below. When storms appear, churning undersea mists take flight and further conceal the presence and intentions of these phantoms. To imagine such a sight—to really focus on the movements, colors, and textures of these Bermudan Bull Sharks—is to enter a special meditation. Take a deep breath...and dive down to that reef.

1910

On New or Insufficiently Described Fishes
Proceedings of the Royal Society of Queensland, 23: 1-55
OGILBY, J.D.

CARCHARIAS SPENCERI sp. nov.

First dorsal inserted a little nearer to the pectoral than to the ventral, its anterior border slightly convex with the outer angle obtusely pointed; posterior angle produced and acute, not nearly reaching to the vertical from the ventral; base of fin rather more than its vertical height. Second dorsal small, inserted a little nearer to the origin of the 1st than to the tip of the tail. Anal originating somewhat behind the 2nd dorsal, its length 1•6 in its distance from the caudal, which is equal to that from the ventral. Caudal long with the upper angle obtusely pointed. Pectoral extending to below or beyond the end of the 1st dorsal, the anterior and posterior borders convex with more or less rounded angles, the outer border emarginated. Space between ventral and anal 1•75 in its distance from the pectoral.

Above ashy or lead blue, below white; tips of 2nd dorsal, caudal, and pectorals darker. (Named for my friend and colleague Adkins Robert Spencer, to whom I am indebted for the specimen above described).

Seas and estuaries of Eastern Australia, ascending rivers to, and it is said even beyond, the furthest limit of tidal influence. attaining a length of 2•5 mm. It is the common "blue shark" of the Brisbane River.

Described from a specimen 122 cm. long, the jaws of which are in the A.F.A.Q. collection; Cat. No. 290.

I very much love the comment, "ascending rivers to, and it is said even beyond, the furthest limit of tidal influence." Regardless of the author's intention, I read this in the lights of grandeur and accomplishment: the shark is literally pushing its conventional saltwater boundaries to their extreme (the tidal limit) only to successfully break them and enter the freshwater realm (the beyond). The incorporation of words such as 'beyond', 'furthest limit', and 'influence'—words commonly associated with journeys, conquests, and discoveries—cultivates a sense of wonder in regards to this species' unique capabilities. Among sharks, this is a type with a commanding power: the ability to break barriers—to be an arm whose reach outstretches all others. While most sharks have less tolerant constitutions, this Bull is able to conquer the salty wall; as a result, it gains precious access to a forbidden realm of untapped resources—namely, the freshwater prey rendered unobtainable by most other sharks. This unique ability makes the Bull strong; not many others can draw from two vastly different ecological realms for resources, and the Bull Shark's power to do so is an incredible evolutionary advantage. Should the seas spoil and rot, the Bull can survive; should the rivers dry and crumble, the Bull *will* survive.

Australia is famous for its riverine Bull Sharks, and it's wonderful that our first Australian entry addresses this subject. I believe that South Africa, the United States (east and gulf coasts), and Nicaragua are the only other nations that share this notoriety, but I could be (happily) proven wrong. In any case, Ogilby here establishes a clear East Australian estuarine and riverine presence, specifically identifying the Brisbane River as a home for this "Blue Shark". I'm assuming the intended common name for *Carcharias spenceri* is "Spencer's Shark", but this synonymy is ultimately of little concern.

1913

The Plagiostomia (Sharks, Skates, and Rays)
Memoirs of the Museum of Comparative Zoology at
Harvard College, 36: 528 p., 77 pl.
GARMAN, S.

CARCHARINUS PLATYODON.

Plate **3**, fig. 4-6.

Squalus platyodon POEY, 1861, Memorias Cuba, **2**, p. 336, pl. 19, f. 5, 6.

Squalus obtusus POEY, 1861, ibid., p. 337, pl. 19, f. 7, 8.

Eulamia obtusa POEY, 1868, Repertorio, **2**, p. 447, pl. 4, f. 1, 2; 1876, An. Soc. Esp. hist. nat., **5**, p. 189.

Carcharias platyodon JORDAN & GILBERT, 1882, Proc. U. S. nat. mus., **5**, p. 243; 1883, Bull. 16, U. S. nat. mus., p. 872, 967.

Carcharinus platyodon JORD. & EVERM., 1896, Bull. 47, U. S. nat. mus., p. 39

Description from a seventeen inch specimen taken at Guadaloupe, W. I.

Reaches a length of ten feet or more.

Common in the Gulf of Mexico and, about the West Indies, in the Caribbean.

There is a big problem here. The specimen that Garman is attempting to describe as *Carcharinus platyodon* is not a Bull Shark. Upon examination of the plate provided, it is very apparent that this seventeen-inch specimen from Guadeloupe is in fact a Sandbar Shark (*Carcharhinus plumbeus*): the first dorsal fin is characteristically large (its apparent size being a trait quite indicative of Sandbars) and the snout is too pointed to fit the Bull Shark's conventional blunt profile.

I didn't include the diagnostics on the opposite page, but instead the references, which provide the primary clues for the reasoning behind this error. If you inspect the two references from the now-infamous (David Starr) Jordan, you will see the inclusion of the common name 'Bull' from the descriptions of *Carcharias platyodon* (1882) and *Carcharinus platyodon* (1896). The first description is most likely correct, but the second is not; in 1896, Jordan also described *Carcharhinus lamia*— which is most definitely a Bull Shark—along with *Carcharinus platyodon*, making the latter a completely different species (probably a Sandbar).

This sort of error is understandable; the world 100 years ago was simply not as technologically connected as it is today, and shark taxonomy was itself young and volatile (being only 150 years old at that time). New species were being discovered every year, and some changed the entire understanding of a taxonomic group (requiring reorganization). Of course, this dynamic is not exactly extinct; new discoveries will always demand new changes, meaning that the entire system is itself unstable (I personally like the word 'entropic', but I'm not exactly sure how appropriate that may be here). In any case, this instability crystallizes the arbitrary nature of names (scientific or otherwise), leading us to ask again: what exactly *IS* the IT?

1913

The Fishes of the Stanford Expedition to Brazil
Stanford University Publications, University Series
STARKS, E.C.

This collection, including the types of the new species, is deposited among the collections of Stanford University. A set of duplicates has been sent to the American Museum in New York.

FAMILY GALEIDÆ.

1.   **Carcharhinus platyodon**   (Poey).

A specimen 29 inches in length, secured at Pará, seems to be referable to this species. The mouth is twice as broad as it is long and the preoral part of the snout is contained $1\frac{1}{2}$ times in the space transversely between the corners of the mouth. The front of the head is semi-circular in outline. The fins are all more or less concave behind. The pectorals when folded back almost reach to opposite the posterior end of the dorsal base. The length of the caudal from the pit on the upper part of its base to its tip is equal to the space between the front of the head and the dorsal fin. The second dorsal is a little in front of the anal, and the base slightly exceeds that of the latter in length. The anal is unlike the second dorsal in form, being deeply notched behind with its lobes almost equal, while the dorsal is concave behind, with its inner lobe reaching far behind its outer.

The color is slaty blue above and pure white below. The dorsal and caudal are outlined in black, especially the caudal, which has a broad, black posterior margin. The tips of the other fins are dark.

I don't have too much to say about this account, other than it's our first confirmation of a Bull Shark presence in eastern Brazil. This *Carcharhinus platyodon* seems to be correctly identified, though we don't have much information to go off of; the coloration, head shape, and fin positioning all seem to be in order, but it would have been helpful to have descriptions of the teeth, eyes, and first dorsal as well.

In any case, this entire dynamic of confusion and misidentification actually serves a greater purpose than to frustrate our own personal interests in order: as *Carcharhinus* is a notoriously difficult genus to demarcate, we can only help but admire the subtle nature of evolution and its responsibility for shaping such closely related but branching forms. From the anchoring root that was the ancestral *Carcharhinus* came a physical archetype that varied across time, space, and demand; key traits such as tooth shape, body mass, and coloration altered as different seas hosted different plays of natural selection. These variations eventually distinguished 33 unique *Carcharhinus* species, but only just, as the same close ancestor remains at the phenotypic core.

Carcharhinus is thus an aesthetic where similarity and distinction duel. It's as if one species 'bleeds through' into another, creating an incredibly consistent spectrum of characteristic phenotypes that, if analyzed alone, would seem to detract from a grander picture. Then again, to distinguish a species is to reflect an incredibly important biological relationship between its members: the exclusive ability to breed (an ultimate distinction of 'kind'). One must reconcile both the holistic and particular identities of *Carcharhinus* in order to fully appreciate its delicately shifting palette; to properly do so, a masterful eye for detail is required. Given this challenging prerequisite, the cause for species misdiagnosis becomes more apparent, understandable, and appreciated.

1929

A Checklist of the Fishes Recorded from Australia Part I

Australian Museum Memoir, 5 (1): 1-144

McCULLOCH, A.R.

Class ELASMOBRANCHII.

Subclass SELACHII.

Order EUSELACHII.

Suborder GALEI.

Family GALEIDÆ.

Genus CARCHARHINUS   *Blainville*,   1816

CARCHARHINUS SPENCERI (*Ogilby*).

1911. *Carcharias spenceri*   Ogilby, Proc. Roy. Soc. Qld. xxiii, 1,

1911, p. 3.   Brisbane R., Queensland.

Queensland.

This is a nice example of early 20th Century Australian shark taxonomy. Most ranks have shifted (with Elasmobranchii and Selachii later becoming a subclass and superorder respectively), while others have been largely abandoned (such as Euselachii, Galei, and Galeidæ). It's interesting to note the synonymy of *Carcharhinus spenceri* and *Carcharhinus leucas* in regards to the origin of such disparity; could specimens from the northern and southern hemispheres not have been compared due to a difficulty in travel? Then again, our Bull Shark's many northern synonyms (which by now are quite familiar to us) have not exactly been resolved at this point in history; perhaps then distance is not the issue, but time (in the sense that more was needed to have accommodated our taxonomic growing pains for this species).

There is absolutely no other piece of new information in the account, so I feel that we should take this time to reflect. Currently, we are standing at the year 1929—or 86 years ago—meaning that we are now less than a century away from our current understanding of *Carcharhinus leucas*. In regards to everything we have seen so far, two primary themes seem most apparent. Firstly, the science here is largely diagnostic: the focus is more so on the species' physical description and less on its behavior or ecological role. Key features that are constantly reoccurring are the Bull's characteristic dentition (triangular, serrated, and slightly curved upper teeth paired with more narrow, broad-based lowers), blunt snout, and concave fins. Secondly, the Bull has a well-established mythos; it is constantly aligned as a dangerous, man-eating species, sometimes even termed 'monster'. I must confess, I find the antiquity of such a reputation to be exhilarating. It's a strange and daunting confirmation of a sort of 'living legend'. Our Bull's modern negativity is not of recent origin: its jaws predate *Jaws*.

1943

Ichthyological Descriptions and Notes
Proceedings of the Linnean Society of NSW, 68: 114-144
WHITLEY, G.P.

GALEOLAMNA (LAMNARIUS) SPENCERI (Ogilby, 1910).

Estuary Shark. "Blue Shark" of Rockhampton and Brisbane districts.

Carcharias spenceri Ogilby, *Proc. Roy. Soc. Qd.*. xxiii, 1910, p. 3.

Brisbane River, Queensland. Type not extant.

Galeolamna spenceri Whitley, *Fish. Aust.*, I, 1940, p. 102, fig. 98; and as
G. (*Ogilamia*) *stevensi*. Ibid., p. 104, fig. 99 (New South Wales
record and figure only, not text).

The species now identified as *spenceri* was figured as *Carcharias brachyurus* by Waite (loc. Cit., 1906), though it disagreed in several respects from Gunther's true *brachyurus* from New Zealand. Waite's specimen, which came from Lake Macquarie, New South Wales, is still in the Australian Museum (Regd. No. I.7586) and may be considered as pleisotype of the species, which was not named when Waite published his figure. Both the late A. R. McCulloch (in MS.) and the writer considered that Waite's example was *stevensi* Ogilby, but neither of us could be sure because it did not exactly agree with Ogilby's description, and moreover, types of neither *stevensi* nor *spenceri* had been preserved. In Queensland, in March, 1943, the present writer made detailed notes on all Whaler Sharks caught, bearing in mind these little-known species of Ogilby, and his present determinations, having been arrived at after most searching comparison, are believed to be correct.

Whitley's notes are fantastically detailed as the author himself, "devised a set of standard measurements for comparative and biometric studies of Australian sharks." There are 49 measurements total, and they are broadly divided into three categories: Head (18), Body (9), and Fins (22). For our "Estuary Shark" specifically, Whitley compared five different juvenile specimens from Queensland and New South Wales, and recorded the measurements for each in a tidy biometrics table. Observantly, the author also made note of the species' lack of an interdorsal ridge, which is today a key trait for distinguishing the Bull form a larger pool of *Carcharhinus* sharks.

Whitley's *Galeolamna spenceri* is without question a Bull Shark due to its unmistakable estuarine habits. However, there is another species in this account that may be of note: *Galeolamna bogimba*, which is today considered to be an old synonym for the Bull Shark as well. The *G. bogimba* as depicted by Whitely is noticeably different from *G. spenceri*. The former is described to have dermal denticles with five separate carinae (or ridges), narrow lower teeth, a proportionally high dorsal fin, and a slightly narrower snout; the latter is described with only 3 carinae, narrow but broad-based lower teeth, a lower dorsal fin, and a broad, blunt snout.

These differences initially led me to favor the idea of a retrospective misdiagnosis, but there is one key piece of information that more or less confirms that *G. spenceri* and *G bogimba* are in fact the same: "*Locality*—Bogimbah, Fraser Island, Queensland." Bogimbah is a creek. This small clue—coupled with the oddity that all five *G. spenceri* specimens were small juveniles and the one specimen of *G. bogimba* was a large adult (weighing 295lbs)—very well supports the idea that Whitely didn't describe two different species, but instead two different stages of the Bull Shark's life.

35

1946

A Descriptive Catalog of the Shore Fishes of Peru
Bulletin of the United States National Museum, 189: 530 pp
HILDEBRAND, S.F.

EULAMIA AZUREUS (Gilbert and Starks)

Carcharias azureus GILBERT and STARKS, 1904, p. 11, pl. 2, fig. 5,

Panama Bay

(original description).—STARKS, 1906, pp. 762, 763, Guayaquil,

Ecuador (compared with type; distribution).

Carcharias milberti MEEK and HILDEBRAND (in part *C. azureus* Glbert

and Starks),

1923, p. 38 (synonymy; description; specimen from New Jersey

compared with one from Guayaquil, Ecuador).

Eulamia azureus BEEBE and TEE-VAN, 1941, p. 109, fig. 18 (range; field

characters; size; references).

Although this species has not been reported from Peru, it may be
expected there, as it has been taken at Guayaquil, Ecuador. A length of 282
cm. has been reported.

Characters for the recognition of this species are included in the
key to the species:

aa.   Snout broadly rounded, its preoral length much less than width of
mouth; teeth, especially in upper jaw, broadly triangular, all distinctly
serrate; origin of second dorsal well in advance of that of anal; no dusky
band on sides

Range.— Costa Rica to Ecuador. Apparently rather common in
Panama Bay.

A Descriptive Catalogue of the Shore Fishes of Peru originates from a 1941 United States Fish and Wildlife Service "mission" described as, "an investigation of the abundance of the fishery resources, the appropriate development of the fisheries, and the proper handling and utilization of the catch." Does this mission statement seem, well, *suspicious* to you? I personally read it with some apprehension, as the United States — my own country — has a well-documented history of neocolonialist tendencies in Latin America. What *exactly* does it mean to investigate Peru's fisheries abundance, "appropriate development" of fisheries, or "the proper handling" of its catch?

This is tricky territory, and I apologize if I'm coming off as a bit strong or alarmist; but the fact is, the United States and Peru have indeed disputed throughout the mid-20th Century over their respective fisheries rights. The former argued for access to marine resources beyond any point 12 miles off the Peruvian coast; the latter supported an exclusively Peruvian jurisdiction stretching to 200-miles offshore (see David C. Loring's "The United States-Peruvian "Fisheries" Dispute." *Stanford Law Review*, Vol. 23, No. 3 (Feb., 1971), pp. 391-453).

In accordance with our modern international Law of the Sea, Peru is of the correct disposition; every coastal country has the right to claim an Exclusive Economic Zone of 200 miles offshore. This renders America's mid-century argument invalid, and that development (when considered with the infamous U.S.-Latin American "resource negotiations" of the United Fruit Company and others) makes the stated purpose of the 1941 U.S. Fish and Wildlife mission a bit unsettling. Don't get me wrong; the science produced by the investigation is solid and insightful. It's just interesting to wonder: what is the likelihood of an ulterior motive behind this American report on Peruvian fishes?

1948

Fishes of the Western North Atlantic
Part I: Lancelets, Cyclostomes, Sharks
Memoirs of the Sears Foundation (for M. R.), 1 (1): 59-576
BIGELOW, H.B. & SCHROEDER, W.C.

Carcharhinus leucas   (Müller and Henle), 1841

Habits. This is a heavy, slow-swimming species, most common inshore in shoal water, perhaps never very far from land except by accident. They are most often caught around docks, at the entrances to the passages between islands, in estuaries and in harbors. They often run up rivers for considerable distances, and it seems that they do not hesitate to enter fresh water. Thus the series we have studied includes one from the Panama Canal at Miraflores Locks, besides others from Lake Yzbal, Guatemala, a body of water that is said to very between fresh and brackish, and a 55-pound specimen has been reported, at least by name, as having been caught in the Atchafalaya River, Louisiana, 160 miles from the sea. We have also received a photograph of a shark four or five feet long that appears to be of this species (unless possibly of the landlocked form *nicaraguensis*, which cannot be determined from the photograph), taken 180 miles up the Patuca River, northeastern Honduras. *C. leucas* is, in fact, the only Shark that is known to have permanently adapted itself to fresh water and developed a local race (see under *C. nicaraguensis*, p. 381). On the other hand, it rarely shows itself at the surface, as the more pelagic members of the genus commonly do, unless lured up by the scent of food, such as floating offal. We have never heard of one jumping, whether at liberty or after being hooked.

Bravo! The famous Bigelow & Schroeder have crafted an absolute masterpiece: *Fishes of the Western North Atlantic* is chockablock with incredible new shark information, and our Bull gets the royal treatment. Hold on tight—a new tide of knowledge is rising!

For the first time in our compendium, a complete description of natural stomach contents is recorded: crabs, smaller sharks, devil rays, porpoises, shad, and mackerel are particularly noted, with the lattermost item being indicative of, "a greater ability to capture fast-swimming fishes than (the Bull Sharks') rather sluggish habits would suggest." The shark is also mentioned to charge stingrays and be undiscriminating in its taste for offal or large bait. Individual sizes of 250, 375, and 400lbs are recorded as well for this species.

The Bull Shark's breeding habits are for the first time addressed, with this account noting, "females with embryos nearly ready for birth have been taken in Florida in October, January and February, which suggests that the young are born there in late winter and early spring". Yet another novelty of this account is the dispute over the deservedness of the Bull Shark's reputation as a man-eater, wherein the author's state, "we think it unlikely that this reputation is deserved, for otherwise shark fatalities probably would be far more frequent than they actually are in Florida and the West Indies, where it is one of the more common of the larger sharks."

I highly recommend perusing *Fishes of the Western North Atlantic* in full, as there is simply too much new and exciting information to compress on this single page. We will doubtless cover the unaddressed topics later, but for now, let us celebrate: Bigelow & Schroeder have brilliantly summarized vagrant pieces of key information into a single, comprehensive, and enlightening account, and for that we are most thankful. Well done!

1950

California Sharks and Rays
California Fish and Game Bulletin, 75: 88pp., 65 figs
ROEDEL, P.M. & RIPLEY, W.E.

GAMBUSO

Carcharhinus azureus

Relationship: A member of the requiem shark family, Carcharhinidae.
Range: Southern California to Ecuador, normally from the Gulf of California south. There is definite record of four specimens caught off Southern California. All were taken in September 1942

Descriptive Characters: Anal fin present; middle of base of first dorsal fin nearer pectorals than pelvics; caudal peduncle without keels; no spiracles; teeth in upper jaw with serrate edges, those in lower finely serrate or fairly smooth; snout short and rounded; width of mouth much greater than length of snout in front of mouth (Figure 16); origin of anal fin well behind origin of second dorsal. **Color:** Gray or blackish above becoming white below; fins tipped with dusky.
Size: Two of the California specimens were over ten feet long.

Importance: None in California. Enters Mexican shark catch at least to a minor degree.
Remarks: This shark has no English vernacular name. Gambuso is given by Walford (1944) as the local name in the Gulf of California.

The California shark fishery of the mid-20th Century receives a healthy treatment in *California Sharks and Rays*. Five different methods of harvest (set gill net, drift gill net, otter trawl, harpoon, and hook and line) are examined in context of the booming economic importance of sharks in the 1940s and 50s. With the mid-40s collapse of European fish oil production (as a result of World War II's general threat to commercial cod fishing's safety), American sharks soared in importance as sources of precious vitamin oil. The large oil-secreting livers of Californian Soupfin and Dogfish sharks were of particular economic interest, and this facet (in combination with the large schooling behaviors of both species) established a very lucrative mid-century market for each shark. Other shark fishing interests include harvesting flesh for dining, meat for baiting, and teeth for curios. Shagreen made from sharkskin (used, as stated by the authors, for sword hilts and cabinetry) is also mentioned as a notable economic product (though one of declining importance at this time).

Though our shark is not mentioned as an important resource for the California shark fishery, it's still interesting to consider its potential as a 'product'. Commercial identity is, oddly enough, a part of the shark's holistic identify; the potential for the Bull to provide palatable flesh, nutritious oil, tough shagreen, and intriguing teeth adds a unique economic dimension to its profile. This dimension in turn summons the idea of 'use', which is a strangely important form of human registration (or, more poetically, human *experience*). The evaluation of the Bull Shark's 'use' requires us to consciously register the shark and incorporate it into our own sphere of existence (either physically in the form of action or mentally in approximation). Thus, we actively *perceive* the shark—we *experience* the shark—and create a unique relationship with the exotic being that it *is*.

1953

Official Common Names of Certain
Marine Fishes of California
California Fish and Game, 39 (2): 251-262
ROEDEL, P.M.

Several rules have governed the selection of official names, all subject to the guiding principle that the purpose of the names is to assist in the compilation of accurate records.

Ideally, a name should be in sole use throughout a species' range and would not be applied to any other North American species. Unfortunately, relatively few names have such universal and sole application. It then becomes necessary to evaluate such names as are applied to a species with regard for the points which follow:

1. Usage in California by both commercial and sport fishermen and the fishing industry.
2. Usage in other areas, particularly adjacent states.
3. Usage on other official lists.
4. The name should not have undesirable connotations which might impair marketability.
5. The name should be brief and descriptive.
6. The name should indicate rather than confuse relationships.

Roundnose Shark _ _ _ _ _ _ _ _ *Carcharhinus azureus* (Gilbert & Starks)
  Listed by Roedel and Ripley (1950) as "gambuso" utilizing a
  Mexican vernacular. The new name is deemed more appropriate
  for use in the United States. This species is very rare in California.

The goal of California Fish and Game is clear and important: establish a single, unalterable, unquestionably perfect common name for each Californian fish species so that an economic and scientific congruency may be reached. The name must be sensible and purposeful: it must, "be brief and descriptive" and "indicate rather than confuse relationships". As such, it should come as no surprise that the name chosen for *Carcharhinus leucas*—in light of its characteristically fearsome reputation, unique tolerance for freshwater, and noticeably heavier body when compared to other members of the genus—is none other than...*Roundnose Shark*?

Truly, the wells of imagination have never been so competently watered. This name is so subversive—so hilariously bereft of every aspiration held by the author—that it has achieved almost nothing in its quest to become a useful mainstay in modern shark study and culture. To be exactly clear about how ironically inefficient 'Roundnose Shark' is as a common name, let us consider the following:

1. We obviously do not call Bull Sharks 'Roundnose Sharks' today, even as an alternative name (such as the less popular but still meaningful 'Cub Shark')

2. We actually don't refer to *any* of the 500 currently known shark species as 'Roundnose Shark' today, even as an alternative name.

3. If we were to run a quick Internet search of the term 'Roundnose Shark', we would pull up an idiotically diverse collection of species bearing no close relationship to each other whatsoever: Basking, Sevengill, Lemon, Great White, and even Sharpnose sharks are all (meaninglessly) valid results.

I apologize for being a bit harsh, but I have a love for the ridiculous. In any case, it must be noted that this account is a sea of useful titles wherein absurdities rarely swim.

1958

Review of the Eastern Pacific Sharks of the Genus
Carcharhinus, with a Redescription of C. malpeloensis
(Fowler) and California records of C. remotus (Dumeril)
California Fish and Game, 44 (2): 137-159
ROSENBLATT, R.H. & BALDWIN, W.J.

Carcharhinus azureus Gilbert and Starks, 1904, p. 11. pl. 2 (Panama).

Early embryos of this species differ considerably from adults and will probably not fit the key. In a 310-mm. specimen the snout is longer and more pointed, the preoral length being contained 1.3 times in the mouth width. The general configuration of the snout is, however, quite similar to that of the adult. A posterior extension of the color of the upper side onto the lower side, as in *C. limbatus*, is also present in this specimen. As is typical for embryonic *Carcharhinus*, the fins are much shorter than in the adult.

The blunt snout and the position of the second dorsal and anal origins set this species off from all eastern Pacific *Carcharhinus* except *C. longimanus*. It is distinguished from *longimanus* in the discussion of that species.

Except for the presence of a mid-dorsal ridge, *C. azureus* appears to be very similar to the freshwater species *C. nicaraguensis*, which in turn is doubtfully distinct from the Atlantic *C. leucas*. The relationships of these species cannot be clarified until more material becomes available. This species has been figured by Gilbert and Starks (1904).

Range: Southern California to Guayaquil, Ecuador.

Material Examined: Mexico, Baja California, vicinity Bahia Magdalena.

The dawn of synonymy is breaking: the idea that *Carcharhinus leucas*, *Carcharhinus nicaraguensis*, and *Carcharhinus azureus* are all the same species has gained new ground thanks to Rosenblatt & Baldwin. Although this speculation is not new, the authors seem to be the first to suggest the need for investigation, importantly stating, "the relationships of these species cannot be clarified until more material becomes available". This implication—taken with the prior professional suspicions about the relationships between the Bull, Nicaragua, and Roundnose sharks—brings more gravity to the concept of synonymy, which in turn inspires more research on the subject. As such, Rosenblatt & Baldwin's report might be part of the first crack in the foundations of incorrect binomials, signaling the eventual eclipse of such names by the currently valid *Carcharhinus leucas*.

Apart from this exciting development, Rosenblatt & Baldwin also offer a detailed explanation of what exactly constitutes the genus *Carcharhinus*. According to the authors, *Carcharhinus* is, "readily separable from all other genera of carcharhinid sharks by the following combination of characters: second dorsal and anal bases about equal in length; second dorsal much smaller than the first; first dorsal inserted close to inner angle of pectoral; labial furrow of upper jaw very short; spiracles lacking; teeth in upper jaw about 30 in number, cusps serrate."

It's funny that we've never come across such a useful set of delineations before. A shark's features become more striking—more brilliant, even—if they are seen with a genus background, as understanding a genus' physical foundations illuminates any evolutionary changes to that basic structure. As such, to understand *Carcharhinus* is to sharpen the definition of *leucas*, and thereby enrich our perception of the species' totality.

1961

Shark Attack off the East Coast of South Africa
1ˢᵗ February, 1961
Investigational Report ORI, 5: 1-4
D'AUBREY, J.D. & DAVIES, D.H.

It is likely that the shark bit the victim at least 4 times. No teeth or tooth fragments were found in the wounds and although the tibia of the right leg was scratched the toothmarks were not characteristic of any particular species of shark.

The flesh wounds had clean-cut edges similar to those of Petrus Sithole who was fatally attacked by a Zambezi River Shark *Carcharinus zambezensis* Peters at Margate, South Africa, in December 1960, Davies and D'Aubrey (1961). The skin lesions made by individual teeth were slit-like indicating that the wounds were caused by flattened *Carcharinid* teeth rather than the prong-like teeth of *Carcharias*. As other sharks with flattened teeth are seldom found in such shallow water it is likely that a species of *Carcharinus* was responsible for the attack.

The size of the sweep of the jaws of the attacking shark was indicated by the wound on the left leg and was $10\frac{1}{2}$ inches at its greatest width.

CONCLUSION

C.zambezensis is a common inshore species along the east coast of South Africa and has been proved to attack man. It is considered likely that the species responsible for the attack on Zimmerman was *C. zambezensis*.

C. zambezensis is a wide-mouthed species and the jaw size indicated by Zimmerman's wounds would have been caused by a specimen of approximate total length 7 feet and weight in the vicinity of 200lbs.

Our first professional report of a Bull Shark attack is horrific. The victim was a 14-year old boy named Geoffrey Zimmerman (5ft 2in, 150lbs, light skinned, wearing black-and-white diamond-patterned swimming trunks at the time of attack). According to the report, Geoffrey entered the cooler waters of Nahoon Beach, East London with friends on an overcast summer afternoon. He had a prior wound—a nail removal from the third tow on his left foot—and was bleeding slightly in the clear but mildly choppy water. At about 2:50 p.m. in a location 60 yards from shore (and at a depth of only 4 feet) Geoffrey started shouting while the waters simultaneously, "became disturbed and red with blood."

As his friends rushed to his aid, they observed a shark make two different advances towards Geoffrey: at the first advance, only the dorsal fin was seen, while at the second, more of the body was revealed as, "the shark seemed to roll on to its side." Though Geoffrey was successfully pulled to shore, he had already endured clean-cut lacerations on his legs and feet (the tibia of the right leg was exposed for about 2 inches), as well as his left arm. On the beach, Geoffrey was, "apparently unconscious". After 20 minutes, an ambulance arrived, but the boy died en route to the hospital.

This is the first shark attack recorded in East London, though the Natal South Coast (an area by this time infamous for 45 documented shark attacks since 1940) lies a couple hundred miles north. Based on the style of the bites, the authors concluded that the culprit was most likely *Carcharinus zambezensis*—the Zambezi River Shark—which nowadays is proven to be yet another synonym for the Bull Shark, *Carcharhinus leucas*. It's strangely terrifying to realize that the Zambezi, Lake Nicaragua, and Australian Estuary sharks—each notorious in their own right—are in fact just one species; I can't rationally explain it, but it feels almost *sinister* that these three 'maneaters' have been Bull Sharks all along.

1961

Shark Attack off the East Coast of South Africa

22nd January, 1961

Investigational Report ORI, 4: 1-5

DAVIES, D.H.

A shark attack took place off the Natal South Coast on January 22, 1961. This is the fourth attack since April 1960 when Michael Hely received serious injuries Davies (1960), Campbell *et al* (1960). At the present rate of incidence of shark attack on humans, this section of the east coast of South Africa is one of the most seriously affected areas in the world.

CIRCUMSTANCES OF ATTACK

The victim, Michael Murphy, was a schoolboy aged 15 years. The colour of his skin was light and he was wearing swimming trunks coloured red and white.

Murphy was attacked at 2.20 p.m. while swimming approximately 30 feet from the shore in water 6 feet deep. The temperature of the water was 26.1°C and there was extensive discolouration due to flooding of silt-laden rivers after recent rains. The weather at the time was bright and sunny with no wind and the tide was low and incoming.

CONCLUSION

The characteristics of Murphy's wounds were very similar to those of Petrus Sithole, the African who was fatally injured by a Zambezi River Shark *Carcharinus zambezensis* Peters at Margate in December 1960 Davies and D'Aubrey (1960). It is therefore considered to be highly probably that the species of shark responsible for the attack on Murphy was *Carcharinus zambezensis*.

The probability of being attacked by a shark is very small. You know this fact; actually, you've probably been bludgeoned over the head (repeatedly) by this fact until the point has been rendered dull. We all know why such trivia exists: so as to spread statistical awareness and ensure that sharks are not 'misunderstood'. But there's a very big problem with this convenient little sound bite: it unwisely dismisses fear.

Regardless of the probability of an attack, the *possibility* is terrifying, and this fear is absolutely valid. Four types of sharks (thirty if you wish to be incredibly inclusive) are known to approach man out of his element and seize him—unprovoked—with mouths full of teeth as sharp as razorblades (literally, not figuratively). As evident in here and the prior South African shark attack record, the damage can be horrifically devastating: though this account's victim, Michael Murphy, lived, his left leg had to be amputated.

I believe I have emphasized this before, but fear is a type of awareness: it's an acknowledgement of power, and the power to kill or maim a human is what the Bull, Tiger, Great White, and Oceanic Whitetip sharks possess. It's part of their identity, and that one little phrase of, "shark attacks are rarer than winning the lottery" dulls this identity down in a very consequential way: it attempts to quell fear—to create a convenience in regards to a wild element—and this is a dismissive mistake.

Understand, I'm not trying to romanticize the damage these species are popularly known to do; I'm only trying to respect these four sharks—to neither demonize *nor underestimate* them. I fear that the "shark attacks are rare" narrative strangely downplays the capabilities of such foreign powers: these four sharks are neither evil nor predictable, and they certainly don't serve your personal narrative. They simply *are* in world that is not yours; they are neither pets nor monsters, but beings with a grave and alien power that should command our respect, not indifference.

1961

*Shark Attack off the East Coast of South Africa 24
December 1960, with Notes on the Species of Shark
Responsible for the Attack*
**Investigational Report ORI, 2: 1-8
DAVIES, D.H. & D'AUBREY, J.D.**

DESCRIPTION OF ATTACK

The victim was seen to be swimming vigorously when suddenly he
screamed and, with his arms flailing, the upper half of his body came
vertically out of the water. His cries ceased and he was seen to fall forward.
The surrounding water became red with blood but the discolouration
rapidly disappeared. Two Africans swimming on the sandbank beyond the
victim swam ashore recrossing the channel without attempting to help him.
Waves washed him in towards the shore and a man who had been standing
on the beach ran in waist-deep and dragged him out.

On being brought to the beach it was seen that both legs had been
severed and bleeding had already ceased. Artificial respiration was given
immediately, and the victim was taken to the Port Shepstone Hospital. It is
likely that he was already dead when he reached the beach.

STUDY OF TEETH FOUND IN WOUNDS

A careful study of tooth fragments and direct comparison between the
two fragments and entire teeth in the collection of jaws held by the
Oceanographic Research Inst., Durban, enabled an identification to be made
of the species of shark responsible for the attack, using a provisional name.

This species has been provisionally identified as *Carcharinus
zambezensis* Peters.

The size and shape of the teeth featured in this account where compared to those of the largest Bull Shark in Durban's Oceanographic Research Institute collection: a 9-foot, 5-inch specimen weighing in at about 903 pounds. Upon examination, it was determined that the attacker was larger than this gigantic specimen.

The investigation incorporated into this account is a very important development in both local and global understanding of Bull Sharks. A survey of sharks living in the Natal region was conducted only a year prior, and the information yielded by over 200 collected specimens implicated that the taxonomy of such sharks was, "in need of extensive revision". As such, the shark responsible for this particular attack—though identified as *Carcharinus zambezensis*—was actually professionally referred to as Species A: an animal bearing very close similarities to *Carcharinus zambezensis*, *Carcharinus nicaraguensis*, *Carcharinus leucas*, and *Carcharinus vanrooyeni*. Like these four other sharks, Species A is, "frequently encountered in shallow, doscoloured water in the vicinity of river mouths and estuaries and is characterised by a heavy blunt-nosed appearance, wide mouth, small eyes and great aggressiveness."

Intelligently, it did not escape Davies & D'Aubrey that Species A could in fact be conspecific with one or all of these four sharks, noting that, "all these species are closely associated with the land and fresh water. With the exception of *C.nicaraguensis* which is a landlocked freshwater species, they are all found close inshore near river mouths and often in estuaries. *C.leucas* and *C.zambezensis* have been found considerable distances up rivers." For the first time in our Compendium, a reference has also been made to the critically endangered Ganges Shark, *Glyphis* (here, *Carcharinus*) *gangeticus*, though its features were deemed too distinct to be included in the candidate pool for Species A.

1963

Preliminary Investigations on the Hearing of Sharks
Investigational Report ORI, 7: 1-11
DAVIES, D.H. & LOCHNER, J.P.A. & SMITH, E.D.

A NUMBER of sharks, including *C. obscurus*, *C.maculipinnis*, *C.leucas* and *S. lewini* were conditioned to respond to pure tones and octave bands of random noise and their thresholds of hearing were determined for these stimuli. The results obtained suggest that sharks are not able to discriminate between frequencies and that their recognition of signals is based on the amplitude versus time characteristics of these signals. They can apparently determine the direction of a sound source accurately.

The present work on the hearing of sharks is being carried out at the Oceanographic Research Institute in Durban as part of a programme of investigation into the biology and behavior of sharks with the ultimate object of devising means of protecting humans from shark attack.

Very little is known about the hearing ability of sharks and it was considered that the most reasonable means for learning more about their hearing was to condition them to respond to sound and to analyse their response to different types of sound stimuli. As far as we know, there are four precedents for the conditioning of sharks to sound stimuli. Vilstrup (1) conditioned Spiny Dogfish (*Squalus acanthias*) to a motor horn, Moulton (2) demonstrated the ability of the Smooth Dogfish (*Mustelus canis*) to associate an oscillator signal with an electric shock, Clark (3) showed that an operant response could be elicited from large Lemon Sharks (*Negaprion brevirostris*) and Kritzler and Wood (4) obtained a provisional audiogram for the Bull Shark (*Carcharinus leucas*).

What is most fascinating about this study is the speculation that sharks cannot discriminate between frequencies, or differences in pitch (for any fellow non-physics-minded readers, these terms relate to the 'lowness' or 'highness' of a sound, but not its volume; think of a piano with a low C and a high C, where the notes can be the same volume, but different lowness or highness, or 'pitch'). Instead, they were deemed to focus on amplitude versus time characteristics, which in simpler terms relates to 'volume' (rather, the 'intensity' of the sound and the changes of such intensity over time). As put by the authors, "the hearing mechanism of the shark is a simple amplitude detector" where the system itself is thought to be, "a velocity-operated mechanism".

Sharks live in a medium where sound travels rapidly—much more quickly than in air. If a sound is made in the ocean, it briskly passes through water particles and causes them to vibrate with a characteristic amplitude and direction. This vibration can be 'matched' by a shark's cartilaginous auditory capsule (which, as speculated by the authors, vibrates with the same amplitude and direction). Inside the capsule, a fluid-filled feature known as the *sacculus* provides a higher impendance to the vibration, and subsequently generates a proportional pressure on the shark's auditory membrane. This establishes a chain reaction where, "pressure is converted into electrical potentials by the receptor cells which in turn trigger the action potentials which are transmitted by the auditory nerve to the brain". It is this transmission that forms the 'sense' of sound.

If a shark is swimming towards a sound source and turns its head right, the intensity of the sound increases in its left ear, and vice versa. As such, the authors speculate that a shark's characteristic swimming motion of 'scanning' left-to right should, "greatly facilitate its ability to follow a sound source". How absolutely fascinating!

1964

Preliminary Guide to the Sharks Found off the
East Coast of South Africa
Investigational Report Oceanographic Research Institute,
8: 1-91
D'AUBREY, J.D.

21. CARCHARHINUS LEUCAS

Common Names:

> ZAMBEZI SHARK, Shovelnose Grey, Slipway Grey (Durban),
> Bull Shark, Cub Shark, Ground Shark, River Shark, Lake
> Nicaragua Shark, Van Rooyen's Shark, Square-nose Shark.

Locality: Specimens longer than 5 feet are common off Durban while smaller specimens are caught at St. Lucia Estuary. It has been caught off the east coast of Africa from the Zambezi River to Algoa Bay and has been reported from Knysna. Small specimens have been caught over 300 miles from the sea in the Zambezi River.

Season: Specimens are caught off Durban throughout the year, the catches being higher between December and March.

Development: Viviparous—the unborn young absorb nutriment from their mother by means of a yolk sac placenta. There are 5 or 6 young in a litter and they are born at a little more than 24 inches in length.

Habits: Feeds on fish, other sharks, rays, squid, and is a scavenger feeding on dead animals and whale meat washed from the shore. An aggressive species, has been proved responsible for an attack on a human off the Natal South Coast and is probably responsible for most of the attacks in this area.

The proper scientific name of *Carcharhinus leucas* seems to be now, at last, firmly established; and yet, D'Aubrey provides the heftiest record of common synonyms yet, proving that there is still so much room for such a colorful collection of titles. As they are not intended to be taxonomic, I find titles as varying as Zambezi Shark to Van Rooyen's Shark more informative than frustrating: they don't truly confuse relationships, but rather enrich our cultural understanding of the species.

Each name reveals a unique cultural niche that this species has attained across various parts of the world, and seeing patterns in such varying niches may hint at certain fundamentals about the animal—its cultural essence, in a sense. The common name patterns for our shark reveal the following as important keys to its human cultural identity: **Freshwater** (Zambezi Shark, River Shark, Lake Nicaragua Shark), **Stocky** (Bull Shark, Cub Shark), **Grey** (Shovelnose Grey, Slipway Grey), **Broad Snout** (Shovelnose Grey, Square-nose Shark), **Cosmopolitan** (Zambezi Shark, Lake Nicaragua Shark, Van Rooyen's Shark).

Shifting back to a biological focus, let's consider viviparity, or "live birth". Bull Sharks are curiously akin to humans in this shared developmental process, in which embryos develop within a placenta directly attached to the mother's body (as opposed to a deposited egg, or even an egg retained inside the mother as in ovoviviparity). Not all sharks exhibit placental viviparity, and the process is in itself a brilliant demonstration of convergent evolution (where mammal and shark viviparity are more or less superficially similar, having evolved independently). Nevertheless, that parallel between Bull Shark and human mothers is incredibly profound, as we can relate on one of the world's most powerful intimacies: the direct nutritional connection between mother and embryo—the 'sharing of life'.

1964

A Survey of Vertebral Numbers in Sharks
Proceedings of the US National Museum, 116: 73-96
SPRINGER, V.G. & GARRICK, J.A.F.

This paper broadly surveys vertebral numbers in sharks. The study
was prompted by our discovery that vertebral numbers are important
systematic characters in those carcharinid shark genera that we have been
investigating (Springer, 1964; Garrick, in ms). We, therefore, have
undertaken to determine if vertebral numbers are of similar value in other
genera, with the hope that some contribution might be made to shark
classification as a whole.

Vertebral numbers have not been used previously as a systematic
character in sharks although they have received some attention in rays
(Ishiyama, 1958) and have been widely employed in teleosts (e.g., Bailey
and Gosline, 1955; Schmidt, 1917). The vertebral numbers from sharks that
have been recorded in the literature are given either without comment or
comparison or are employed as data for studies in morphology or
intraspecific variation (Punnett, 1904; Aasen, 1961).

We present here vertebral data on 1524 specimens. We personally
made counts from 858 of these, mostly by X-ray methods. The remaining
666 counts are from the literature or were supplied by colleagues. The 1524
specimens pertain to 70 of the approximately 80 genera of sharks and to
135 of the approximately 300 species. Because our purpose has been to
survey, our coverage within individual species is far from complete, but
considerable attention has been paid to a few species that have presented
problems.

Sharks as a whole range in their total vertebral counts (T) from about 60 to 400 (with the average number of vertebrae being around 150); considering this and the fact that 1524 specimens were sampled, we could expect Springer & Garrick to have examined about 228,600 vertebrae — over a quarter of a million. This incredible endeavor was an important step in anchoring vertebral counts as a useful taxonomic tool, as it yielded new insight on relationships and distinctions that could not have been accomplished through external examination alone. One of the more interesting examples of this would be the findings with *Squalus acanthias*: dogfish from the North Atlantic were found to have 79-85 precaudal vertebrae while dogfish from the North Pacific were found to have 68-76 (indicating that the species had actually differentiated). Vertebral counts are still used today as a key identifier of physically similar but truthfully distinct species (one of the most famous modern examples being the delineation of the Carolina Hammerhead *Sphyrna gilberti* from the Scalloped Hammerhead *Sphyrna lewini*).

In terms of *Carcharhinus leucas*, there's comparatively little dramatic insight. Five measurements were taken:

P	Precaudal vertebral count	110-114
C	Caudal vertebral count	95-104
T	Total vertebral count	208-218
A	$\dfrac{Length\ of\ penultimate\ monospondylous\ centrum}{Length\ of\ first\ dispondylous\ centrum} \times 100$	110-130
B	$\dfrac{Length\ of\ penultimate\ monospondylous\ centrum}{Diameter\ of\ penultimate\ monospondylous\ centrum} \times 100$	60-72

These numbers were taken from 7 specimens from North and Central America.

1967

Caligoid[s] Parasitic on Sharks in the Indian Ocean
Proceedings of the US National Museum,
121 (3572): 1-21, figs. 1-54
CRESSEY, R.F.

During the International Indian Ocean Expedition, 35 species of caligoid copepods were collected from 29 species of sharks. The author collected during Cruse 5 of the R.V. *Anton Bruun* and also at Nosy Bè, Madagascar. In addition, several other participants collected copepods and donated the material to the author for inclusion in this study.

Of the 35 species of copepods collected, 8 of them were new. Four of these new species, all members of the Eudactylinidae, are described here. Two of the others represented new genera (*Pagina* and *Bariaka*) and each has been described separately elsewhere. The remaining 2 new species are members of the Pandaridae and are being described in a paper currently in press revising the entire family.

A map of the Indian Ocean showing points of collection is included (fig. 54). All station numbers refer to points on the various cruise tracks of the R.V. *Anton Bruun*. In text tabulations, Roman numerals refer to spines, Arabic to setae.

I would like to thank the following persons for their efforts on my behalf: Dr. Richard Gooding, University of Singapore; Miss Sherril Kite, Woods Hole; Dr. Alan Lewis, University of British Columbia; Mr. Richard Shomura, Bureau of Commercial Fisheries; Dr. Marta Vannucci, University of São Paulo; and collections from sharks caught near Durban, South Africa, that I have included in this paper.

Parasites carry a uniquely visceral form of disgust when they come to mind; it's an interesting gravity with roots deeply seated in biology, as our bodies and minds are evolutionarily attuned to harmful things—things that would literally sap our energy. These things are grimy, slimy, digging, disgusting, mooching, loafing, burrowing, gnawing, biting, sucking, sapping, creeping, crawling, simply awful perversions of our self-regulating, energy-conservative structure, and with that they have incurred one of our most sincere forms of hate, fear, and aversion. However, this could blind us to the brilliance of parasitism.

Finding a larger, permanent store of energy and aligning your entire existence to that energy is simply genius. Parasites are often extremely specialized in their adaptations to better coexist with their host, and this gives rise to rather glittering array of host-specific forms. From Florida, Madagascar, and South Africa, eight parasites have been collected here, each exploiting a unique niche within their home of *Carcharhinus leucas*.

Species Name	Parasitic Habitat
Pandarus carcharini	Body surface and fins
Perissopus dentatus	Caudal fin and right clasper
Nesippus orientalis	Roof of mouth and gill arches
Nesippus cyrpturus	Roof of mouth and gill arches
Alebion gracilis	Body surface and fins (dorsal surface usually)
Paralebion elongatus	Mouth and gill arches
Nemesis robusta	Gill filaments
Kroyeria gracilis	N/A

1967

Revision of the Family Pandaridae (Copepoda: Caligoida)
Proceedings of the United States National Museum
121 (3570): 1-133, figs. 1-356
CRESSEY, R.F.

In 1907, C. B. Wilson published a revision of the subfamily
Pandarinae as part of a series of papers dealing with caligoid copepods. We
now recognize that much of this work was superficial, containing
descriptions of species often incomplete and inadequately figured;
nevertheless, it served to focus attention on a group of parasites,
caligoid copepods, which were then and are still today poorly known in
most cases.

Between 1960 and 1965 I collected and solicited material of the
family Pandaridae from as many different areas as possible. As a result of
this accumulation of material and data, I feel that a revision of the family is
in order. Ecological relationships are now more evident than before.

Because of inadequate species descriptions that exist for most
members of this family, positive identification of material is often difficult.
This results in the publication of records that obscure our understanding of
existing host-parasite relationships. I believe that, in most cases, I have
examined enough samples of members of this group to be able to draw a
clearer picture. Also, as a result of these collections, I have been able better
to define important taxonomic characters and to discount others on which
new species descriptions have often been based. It is with the foregoing in
mind that I have made the following family revision. The Pandaridae as
defined here is composed of 12 genera and 33 species.

Five species here are said to call *Carcharhinus leucas* home. What a fantastic relationship: that individuals of a completely separate, foreign taxon come to the treat the shark as a sort of planet, depending on it almost exclusively for resources, protection, and living space. Though unbeneficial to the shark itself, this relationship is a rather overlooked poetry; the host is a life-giver, a patron—a foundation on which others depend. It's a profound position, and perhaps an unorthodox understanding in regards to our modern shark culture: along with being an apex predator responsible for death, the Bull Shark is at times a vulnerable victim and propagator of life.

Scientifically speaking, the five Pandarid copepods noted to parasitize *Carcharhinus leucas* are each very well adapted to exploit a specific ecological niche found within their host. *Perissopus dentatus* is unique among the others in its ability to cement itself to its preferred location around the nares (which are often bumped as a result of the shark's investigations); such a chaotic place requires a secure holdfast, and the cementing maxilliped of *Perissopus* does an excellent job as such an anchor. The two species of *Nesippus* (*orientalis* and *crypturus*) each possess maxillipeds with sharply pointed, hook-like tips ideal for digging into their preferred habitats of the shark's gill arches and roof of the mouth. The *Pandarus* species (*sinuatus* and *carcharini*) also have hook-like maxillipeds, but with tips of a more spatulate design so as to better grip the shark's dermal denticles on the outer body surface and fins.

Regardless of ecological niche, all five of these species of Panadarid copepods share a pronounced sexual dimorphism: females are generally more compact in build than males, and are adapted to a more sedentary lifestyle. Males are conversely more mobile, and have strong maxillipeds that may help them hold on to females during copulation.

1970

Copepods Parasitic on Sharks from the West Coast of FL
Smithsonian Contributions to Zoology, 38: 1-30, figs. 1-110
CRESSEY, R.F.

Between 1965 and 1970 over 400 collections of parasitic copepods representing 31 species were taken from 16 species of sharks, mostly from off Sarasota, Florida. The work was accomplished in cooperation with the Mote Marine Laboratory (formerly Cape Haze Marine Laboratory) while the laboratory was engaged in activities concerning shark biology. Larger species of sharks were caught on trot lines set from one to five miles off Siesta Key. Smaller species were caught by using a modified trot line set closer to shore off Siesta Key and also in various locations in Tampa Bay.

This paper represents an account of those species of parasites present on the 16 species of sharks collected. Five new species are described. No attempt has been made, in most cases, at redescribing the known species since it is my intention to do this later in revisionary works concerning several of the genera represented here. The large collection reported here, together with collections from other parts of the world, will enable me to more effectively do this revisionary work. *Paralebion elngatus* Wilson, however, has been redescribed since it does not fit into any proposed future revisionary work.

The material representing four species of *Alebion* reported here has enabled me to demonstrate the usefulness of the spermatophore as a taxonomic character in this genus. The collection also adds much to the information previously known regarding the host specificity of many of the copepods. All figures were drawn with the aid of a camera lucida.

This report of Cressey's describes the following species as parasitic of Bull Sharks (the number of samples taken for this study are indicated within the parentheses): *Alebion carchariae* (11), *Paralebion elongatus* (25), *Nesippus orientalis* (26), *Nesippus crypturus* (13), *Perissopus dentatus* (17), *Pandarus smithii* (1), *Pandarus cranchii* (1), *Pandarus sinuatus* (10), *Kroyeria spatulata* (9), and *Nemesis atlantica* (14).

For some species, Cressey provides new insight into the copepods' abundance, season, and habitat preferences. Species deemed 'common' include *A. carchariae*, *N. orientalis*, *N. crypturus*, and *K. spatulata*. Many species showed no true seasonal preference (*P. dentatus*, *P. sinuatus*, and *N. atlantica*), though *Alebion carchariae* is largely absent in the summer, opting instead to parasitize on Florida's winter inhabitants. Interestingly, *Nesippus orientalis* and *Nesippus crypturus* were often observed together within the mouth of their host (the former often observed in clusters). *K. spatulata* and *N. atlantica* found shelter within the shark's gills while *P. sinuatus* preferred the outer body surface and *P. dentatus* was, "most often found attached near the external nares or on the trailing edges of the fins."

To examine these parasites is to embrace an incredibly detailed, alien world hidden within our midst; they're right in front of us, attaching themselves to one of the planet's most charismatic creatures, and yet, they are overlooked. Is it size that averts our eyes? Or are they simply too alien—to disturbingly removed from our common perceptions of life and biodiversity? Whatever the reason, these amazingly varied masters of manipulation are not given due recognition.

Skillful, resourceful, and perfectly designed for their powerful and challenging habitat: these are the qualities that define the parasitic copepods.

1973

Sharks of the East Coast of Southern Africa
I: The Genus Carcharinus (Carcharhinidae)
Investigational Report ORI, 33: 1-168
BASS, A.J. & D'AUBREY, J.D. & KISTNASAMY, N.

This paper, the first of a series on the sharks of the east coast of southern Africa, reviews the genus *Carcharhinus* as recorded in this region. Research on these animals was motivated by a relatively high incidence of shark attack. For instance, forty-eight attacks were recorded from South African seas between 1940 and 1958. All but seven of these took place on the Natal coast, an area where the economy is firmly based on tourism. Although the number of injuries and deaths due to sharks is negligible when compared to that caused by other factors such as road accidents, the psychological impact of a shark attack is such that a single attack can seriously affect the economy of a coastal resort. A great deal of effort has therefore gone into finding methods of preventing shark attack. The first attempt at protection in South African seas was made in 1907 when a bathing enclosure was erected in Durban. This lasted until 1928 after which Natal beaches remained unprotected until 1952 when a netting system was introduced in Durban. During the previous nine years a total of 21 attacks, seven of which were fatal, had taken place off Durban beaches but in the two decades since the introduction of meshing there have been only two minor incidents. Unfortunately, practical difficulties in laying and servicing nets kept other resorts from installing similar systems and in 1957/1958 seven attacks took place along the southern Natal coast resulting in a severe recession in the economy of the area.

Like Bigelow & Schroeder's *Fishes of the Western North Atlantic*, this account by Bass et al. is incredibly detailed and compellingly saturated with knowledge on the South African Bull Shark and its kin across the globe. *C. zambezensis, C. azureus, C. spenceri, C. bogimba, C. greyi mckaili, C. nicaraguensis,* and *C. vanrooyeni* are all once again consolidated into the now-unquestionably correct binomial of *Carcharhinus leucas*, but there is one final outlier: the sinisterly similar doppelganger, *Carcharhinus amboeinensis*. This shark—known colloquially as the Pigeye Shark—differs from the Bull in only that it has fever precaudal vertebrae and a slightly taller first dorsal fin.

Life stages and size classes are for the first time clearly defined in our Compendium. Neonatal sharks range from 60-70cm, and are born with a characteristically juvenile coloration of black fin tips and fringes (these begin to fade at around 100cm and are fully displaced by the adult coloration at around 200cm). Adolescents of the Natal Coast are described to be those ranging from 150 to 230cm, with maturity beginning at roughly 225cm. Males fully mature at 250cm while females do so at 260cm. The largest reported specimen is here described to be 300cm.

Incredibly, *Carcharhinus leucas* was recorded in St. Lucia Lake habitats with 53ppt water (about 50% saltier than normal seawater, the saltiness being induced by drought). This brine tolerance confirms a salinity range of about 50ppt for the Bull Shark (with 0ppt being freshwater, 35ppt seawater, and over 35ppt brine), which is almost unimaginable; only 1% of the known sharks can even try to tackle such an outstanding range (with *Glyphis* species being the only contenders), but I'd be surprised if they can topple the Bull as Euryhaline King.

Now this...this is *essence*. Take note, reader: I think we found something...something truly, essentialy, irreplicable in all of nature.

1973

Spinal and Cranial Deformities in the Elasmobranchs
Carcharhinus leucas, Squalus acanthias, and
Carcharhinus milberti
Journal of the Elisha Mitchell Sci. Society, 89 (1-2): 74-77
SCHWARTZ, F.J.

A second occurrence of a spinal deformity is reported for a bull shark, *Carcharhinus leucas*, captured in Chesapeake Bay. Deformed snouts and head conditions are also noted for *Squalus acanthias* and *Carcharhinus milberti* from North Carolina waters.

Incidences of spinal or chondocranial deformities are rare in elasmobranchs (Dawson, 1960, 1964, 1971). Springer (1960) mentioned a deformed female *Carcharhinus leucas* that produced six "normal" embryos (Clark. 1964). He also supplied Clark, who reported (1964) a humpback and lateral spine curvature in a *C. leucas* from Sarasota, Florida, a small hammerhead with a mild spinal curvature. Clark (1964) also commented on the capture of a young bonnethead shark with a spinal curvature. Templeman (1965) believed that the spinal deformities that he found in *Raja* were produced during embryonic development. Hickey (1972) reviewed some of the common causes and effects of abnormalities in fishes. This paper records a second spinal deformity in *C. leucas* and deformed shortened snouts in two other sharks, *C. milberti* and *Squalus acanthias*.

Mrs. Macellus, R.N., Calvert County Hospital, Prince Frederick, Maryland, produced the radiographs of *C. leucas*. C. Nicholson, M.D., Morehead City, N.C., provided helpful medical interpretation of the *C. leucas* deformity.

It's strange to imagine an imperfect shark. We terrestrial beings are too isolated from the ocean to attain a true familiarity; we only see glimpses—short images from the surface waters, the beach, or television—and thus only understand a select fraction of the ocean's true from. It's a place of worms, disease, and disorders—a world not so unlike our own in regards to the biological problems its denizens may face. Sharks in particular have a constant cultural image of 'perfection' or 'impeccable design' (and no doubt, I have contributed to that culture), but the fact is, they too are subject to anomaly.

Miscalculations in embryonic development often result in death, either immediately or as a result of eventual physiological failure (such as an organ complication that worsens with age, or a deformity that inhibits the acquisition of food). However, the malformed individual that Schwartz describes has attained a reasonable size of 114.3 kg, meaning that its ailments do not appear to be incapacitating. This Bull Shark has a deformed spine resulting from a combination of ankhylosis, lordosis, and scoliosis. These conditions cause some vertebrae to fuse and callus to develop, resulting in an abnormal curvature overall (though Schwartz noted, "These conditions were not so severe as noted for some teleosts, such as cods").

Schwartz makes an interesting point that, "Injury, disease, or malnutrition were speculated as the causal agents for the deformity". I personally find it odd that genetics were not mentioned as a possible contributing factor; it seems as if something so structurally fundamental as spinal development could only be vulnerable to deeply seated causes— namely, genetic miscoding. Nevertheless, Schwartz specifically cites tuberculosis and salmonella as possible diseases that could have afflicted the shark early on, as each is equally capable of ultimately corrupting spinal alignment.

1978

Brain Organization in the Cartilaginous Fishes
In: Hodgson, E.S. & Mathewson, R.F. (eds.).
Sensory Biology of Sharks, Skates and Rays.
Office of Naval Research, Washington, D.C.: 117-193
NORTHCUTT, R.G.

Until recently elasmobranchs were considered primitive fish with small, simple brains mediating a behavioral repertoire limited compared to those of bony fish or land vertebrates. The elasmobranch telencephalon was said to function primarily in olfaction, and its efferents were believed to project principally to epithalamic and hypothalamic centers integrating olfactory and gustatory behavior. The roof of the midbrain was believed to be the highest visual center—capable of only crude visual analysis—where ascending somatic and visual sensations were integrated into a few stereotyped behavioral responses. The well-developed cerebellum was believed to relate to powerful, well-coordinated trunk movements, yet sharks were said to be clumsy.

These conclusions are rapidly being replaced by a newer picture of elasmobranch central nervous system (CNS) organization. However, it is important to understand how sharks came to be viewed as primitive, robotlike smelling and feeding machines. Nonmammalian vertebrates were assumed to represent earlier, and thus simpler, stages in the evolution of mammals. Attention focused on...assigning different vertebrates to different "phylogenic levels". Elasmobranchs have cartilaginous skeletons, which were believed to predate bone in vertebrate evolution. Thus, they were assigned to a primitive ("low") position in vertebrate evolution.

Northcutt's *Brain Organization of the Cartilaginous Fishes* appears to be a fundamental in the field of shark neurobiology. The CNS is described in excruciatingly careful detail, while revolutionary conjectures on the intelligence of sharks and their behavioral capacities are brilliantly explored. Contrary to the scientific dogma of the 1960s and early 70s, Northcutt found sharks to be in no way 'small-minded', as he documented brain-to-body ratios, encephalization quotients (EQs), and learning capabilities comparable to those of both birds and mammals. In regards to *Carcharhinus leucas*, Northcutt observed that, "The myliobatiformes and carcharhiniformes are characterized by the most complex neural development among living elasmobranchs" and that, "advanced galeomorph sharks possess relative forebrain development comparable to that of endothermic vertebrates". Carcharhinids, Sphyrnids, Dasyatids, and Mylobatids—families with, "the most complex neural organization and the highest brain-to-body ratios known for elasmobranchs"—all curiously share a unique adaptation: the yolk sac placenta. Carcharhinids were also noted to, "occur widely in reef communities, which are also the habitat of the largest-brained teleosts (Bauchot et al. 1977)." This may be related to a, "selective advantage in predators learning the complex spatial organization of the reef habitat, and...pursuing prey that is well-camouflaged."

In spite of these incredible insights, Nortchutt also cautioned that sharks possess structures, behaviors, and cognition that may not have a parallel with those of mammals (given their 400-million year difference in evolutionary history). As such, any intelligence 'discrepancy' must not simply be misconstrued as 'deficiency' (though it obviously has been in past scientific literature). Unfortunately, the research required for a truly objective assessment of shark intelligence has been, as of 1978, severely lacking.

1981

Requins de Méditerranée et d'Atlantique
(Plus Particulièrement de la Côte Occidentale d'Afrique)
Fauna Tropicale, 21: 1-330
CADENAT, J. & BLACHE, J.

Carcharhinus leucas [(Val.) Müller et Henle, 1841]

Espèce cosmopolite commune, fréquentant les eaux littorals de toutes les mers chaudes, penetrant en estuaires et lagunes et se rencontrant meme en eaux douces (etude générale de ces localisations *in* BOESEMAN, 1964), où elle a reçu parfois un statut spécifique particulier.

Nous avons personnellement observe l'espèce au Sénégal, en Guinée, Sierra Leone et Côte d'Ivoire, au Dahomey et au Congo (où des mâchoires de cette espèce ont été attribuées souvent, par erreur, à *Carcharodon carcharias*);

SPRINGER et GARRICK (1964) ainsi que SADOWSKY (1971) décomptent 109—115 précaudales (M = 113) + 95—104 caudales = 208—218 vertèbres au total, sur des exemplaires de l'Atlantique occidental.

Le rapport hépato-somatique, très variable, peut être assez élevé : de 5,1 à 5,2 % pur des jeunes de moins de 2 m, d'octobre à janvier, jusqu'à 18,1 % chez un mâle adulte en décembre ; chez les fœtus à terme, en juin, le rapport atteignait 17,3 à 18,8 %.

Le régime alimentaire est essentiellement ichthyophage ; ce requin est souvent cause d'importants dégâts aux filets maillants ou chaluts, lorsqu'il vient dévorer les poisons captures lors de la relève de ces filets.

L'espèce est bien connue sous les noms de : bull shark (An), requin taureau (Fr).

The Hepatosomatic Index (HSI) is a value used to measure the metabolic activity of a fish, calculated as:

$$\text{HSI} = \frac{\textit{Weight of the Liver}}{\textit{Weight of the Body}} \times 100$$

Fish with higher HSI values tend to be healthier, and are often found in environments with favorable, energy-rich conditions. This is due to the fact that in an energy-rich environment, a fish tends to consume more, and thus requires a larger liver to process the higher intake in potential energy. HSI is a value directly dependent on liver size; a larger liver accounts for a higher percentage in body mass, and thus results in a higher liver-to-body ratio (in essence, HSI). In energy-poor environments, fish livers tend to be proportionally smaller, lending to a likewise smaller HSI value.

Our first encounter with HSI can be found here in *Requins de Méditerranée et d'Atlantique*, wherein Bull Sharks are described to possess the following values:

STAGE	MONTH	HSI
"Chez les fœtus à terme" (full-term fetus)	Jun.	17.3 to 18.8%
"Des jeunes de moins de 2 m" (juveniles under 2m)	Oct. to Jan.	5.1 to 5.2%
"Un homme adulte" (an adult male)	Dec.	Up to 18.1%

Cadenat and Blache found the HSI of Bull Sharks to be remarkably variable, though capable of becoming quite high. Variation of HSI is partly seasonal, but can also account for discrepancies in water quality; with this latter point in mind, could the strikingly low HSI of juveniles—a class most often associated with freshwater and estuarine systems—be partially derived from a closer proximity to human influence?

1982

Sharks of the Genus Carcharhinus
NOAA Technical Report NMFS, Circular, 445: 1-194
GARRICK, J.A.F.

The genus *Carcharhinus* Blainville contains 25 living species of
whaler sharks, one of which (*C. wheeleri*) is described as new while the
other 24 incorporate 95 identifiable nominal species which fall into the
limits of the genus as here recognized. Features studied include
morphometrics, external morphology, color, tooth numbers and shapes,
vertebral numbers and other vertebral characteristics, and biological data.
The systematic value of these features is reviewed, and it is concluded that
despite their importance at the specific level they do not in general allow
firm statements on subgeneric groupings or on the relationship between
Carcharhinus and other similar genera. Accordingly, no formal
subdivisions of the genus is proposed, and the limits and characterization of
the genus are essentially as in Bigelow and Schroeder (1948) except that the
following six nominal species are excluded because of one or more notably
divergent aspects of their morphology: *Carcharias gangeticus* Müller and
Henle, *C. glyphis* M. and H., *C. oxyrhynchus* M. and H., *C. temminckii* M.
and H., *Carcharhinus tephrodes* Fowler, and *Carcharhinus velox* Gilbert. A
further 13 nominal species are treated as species dubia.

The 25 species are predominantly tropical-subtropical, but only
two appear to be confined to the tropics and seven have been recorded from
the tropics to latitudes as high as 40°. Most are coastal, one is virtually
insular, and one, or perhaps two, enter fresh or brackish water. Eight
species are worldwide.

This NOAA Circular is yet another compilation of all prior information on *Carcharhinus leucas*; much of this data is familiar, but there are a few sparkling details that we have never seen before. *Carcharhinus amboinensis*—the eerily similar Pigeye Shark—is mentioned to have one or two fewer teeth than *C. leucas* in the lower jaw. This characteristic—along with *C. amboinensis'* fewer precaudal vertebrae and higher first dorsal height: second dorsal height ratio—enriches our respect for subtlety. To distinguish *Carcharhinus leucas* from *Carcharhinus amboinensis* is to command a patient and practiced eye. Such a masterful eye may also observe an incredible new color detail on our Bull: tiny blue spots. In living specimens, these spots may be found within the very familiar drab of grey, brown, and yellowish-white on the body—adding an almost fanciful layer of color to how we *understand* this shark.

The incorrect notion that populations of Bull Sharks are landlocked in Lake Nicaragua is dispelled. Tagging studies found that the Lake Nicaraguan Bull Sharks utilize the Rio San Juan to access the Caribbean Sea (though this is an idea we have already seen, it has beforehand only been in the form of speculation). Some slight physical differences between regional populations of *Carcharhinus leucas* are noted as well. The Bull's first dorsal fin usually originates above or just behind the pectoral axil; however, in individuals from the Indian Ocean and Indo-Australian region, the first dorsal tended to originate further back. In concordance with this observation, Western Atlantic Bull Sharks were found to possess fewer total vertebrae than their Indian and Australian counterparts. However, there is an expressed concern for the misinterpretation of data on Indian and Pacific populations of *Carcharhinus leucas* as individuals from that region have been so often confused with *C. amboinensis* and *Glyphis gangeticus*.

1985

Estimated Catches of Large Sharks by US,
Recreational Fishermen in the Atlantic and Gulf of Mexico
In: Shark Catches from Selected Fisheries off the U.S. East
Coast. By Emory D. Anderson, John G. Casey, John J.
Hoey, and W. N. Witzell, NOAA Technical Report, 31: 15-19
CASEY, J.G. & HOEY, J.J.

Several species of large Atlantic sharks are an important resource to the U.S. recreational fishery (Table 1). Sharks have been fished commercially in the past (Springer 1952) and, despite their present low value, the stocks are considered potentially valuable to U.S. commercial interests. World landings of elasmobranch fishes (sharks, skates, rays) in 1981 were 600,607 t (metric tons), or about one-fourth of the world's combined landings of tuna, swordfish, and billfishes (Thompson 1983). In the face of increasing world demand for food and byproducts from the sea, an increase in the harvest of sharks in the U.S. Fishery Conservation Zone is assured.

In the Atlantic, new fisheries for sharks are likely to develop along several lines as the demand for recreational opportunities, and the value of flesh, fins, or byproducts increases. Judging from the recent growth of the recreational fishery for sharks off the U.S. northeast coast and the continuing interest of fishermen in sharks as "big game fish," recreational fishing for large sharks will continue to increase along the entire Atlantic coast. Currently a high percentage of the sharks caught by recreational fishermen are released or discarded with the remainder being mounted for trophies or brought home for food.

	Atlantic North of Virginia	Atlantic South of Virginia	Gulf of Mexico	TOTAL
Survey %	< 0.1%	5.4%	7.0%	2.1%
Tagged %	0.1%	1.9%	4.3%	0.5%
# Of Sharks	NA	3,229	3,248	4,839
Avg. Weight	NA	103 lbs.	103 lbs.	103 lbs.
Sum Weight	NA	332,587 lbs.	334,544 lbs.	498,417 lbs.

The above table summarizes the percentage (relative to other shark species), number, and weight (both individual averages and total landings) of Bull Sharks as recorded by Casey & Hoey in this NMFS report. The authors primarily draw from Hamm and Slater's 1979 marine recreational fishing survey and the NMFS Cooperative Shark Tagging Program (providing the table's Survey and Tagged percentages, respectively). The shark landing data as provided by these studies is sorted into three geographic categories: Atlantic North of Virginia, Atlantic South of Virginia, and Gulf of Mexico.

Considering that the authors reported an estimated 22,600,000 lbs. in total landings of all large shark species in this area, there seems to be a potential viability for an American Atlantic shark fishery. However, there is also a serious concern for increased fishing mortality as sharks grow and reproduce at a characteristically slow rate. Casey & Hoey specifically mention the fisheries crash of the Porbeagle (*Lamna nasus*) in the Western North Atlantic after the mid 1960s as an example (with growth and maturity of the species taking about 6 to 9 years). Dusky, Sandbar, and Sandtiger Sharks have experienced similar declines as a result of Western North Atlantic fishing operations.

1986

Fichas FAO de Identificação de Espécies para Propósitos Comerciais. Guia de Campo para as Espécies Comerciais Marinhas e de Águas Salobras de Angola
FAO, Rome. 184 p.
BIANCHI, G.

SUMÁRIO

Este guia de campo inclui espécies de peixes ósseos, marinhos e de águas salobras, tubarões, raias, lagostas, camarones, caranguejos, estomatópodes, cefalópodes e tartarugas marinhas com interesse comercial em Angola. Cada um destes grupos de recursos é precedido de uma introdução sobre a terminologia técnica utilizada, seguida de uma lista de anotações e ilustrações das espécies mais importantes que inclui os nomes FAO, os nomes locais mais vulgares, o tamanho máximo, o habitat e as artes de pesca utilizadas. No sentido de facilitar a primeira identificação, o grupo "Peixes ósseos" é também precedido de um guia ilustrado das diferentes familias. Um índice composto pelos nomes científicos e vernaculares das famílias e das espécies é também incluido.

Carcharhinus leucas (Valenciennes, in Müller & Henle, 1839)

Sinónimos: Carcharhinus nicaraguensis (Gill & Bransford, 1877)

Nomes FAO: Es – Tiburón sarda; **Fr** – Requín bouledogue; **In** – Bull shark

Tamanho: Máx. 350 cm, comum até 260 cm

Habitat e biologia: Uma espécie costeira, encontrada também em águas salobras e em rios. Vivípara, alimenta-se de peixes e de invertebrados.

Artes de pesca utilizadas: Capturada com linhas e provávelmente com anzol.

Here is our first encounter with the Food and Agriculture Organization of the United Nations (FAO), which has been a historic pillar of shark distributional research and data (perhaps most famed for their 'Sharks of the World' catalogue and similar regional publications, which we will surely see more of later). As this guide is intended for quick in-field reference, each species entry is succinct and offers only the essentials in regards to terminology, size, habitat, and harvest method.

Carcharhinus leucas is thus described with the FAO standard names of Tiburón sarda (Spanish), Requín bouledogue (French), and Bull shark (English), with *Carcharhinus nicaraguensis* being the only synonym. The maximum size attributed to this species is 350 cm, with an average of 260 cm. Characteristically, it is defined as a live-bearing coastal shark capable of entering fresh and brackish waters and feeding on a variety of fish and invertebrates. Hook-and-line is the only method of harvest mentioned.

As with other FAO publications, a primary goal of this field guide is to facilitate the correct identification of commercial species so that accurate statistical counts may be generated for their fishery. Such accuracy is crucial to proper fisheries planning, as misidentification can mistakenly cripple or bolster the effort to manage a particular species. Sharks are especially prone to the risk of misidentification as many species are extremely hard to distinguish from one another in the wild (*Carcharhinus* being one of the more notorious genera to delineate). Furthermore, fishermen may sometimes misapply the names of popular species such as the Great White to sharks of a lower presence in popular culture, thus adding a 'charisma bias' to their counts. Bull Sharks—being both a *Carcharhinus* and an infamously dangerous species—may have a higher sensitivity to this misidentification problem as a result.

1986

Levantamento Faunístico dos Elasmobrânquios
(Pisces: Chondrichthyes) do Litoral Ocidental do
Estado do Maranhão, Brasil

Boletim do Laboratório de Hidrobiologia, São Luis, 7: 27-41

LESSA, R.P.

RESUMO

O levantamento dos peixes elasmobrânquios foi realizado de outubro de 1983 a dezembro de 1985 na costa do Estado do Maranhão, Brasil, entre a Ilha de Santana e a Barra de Lençóis (1° 20' S à 2° 30' S) através de pesca com rede de emalhar de deriva nas diversas baías que compõem a Zona das Reentrâncias Maranhenses.

As seguintes espécies foram capturadas:
Carcharhinus acronotus, Carcharhinus leucas, Carcharhinus limbatus, Carcharhinus porosus, Carcharhinus perezei, Isogomphodon oxyrhyncus, Rhizoprionodon lalandii, Rhizoprionodon porosus, Ginglymastonia cirratum, Sphyrna lewini, Sphyrna tiburo, Sphyrna tudes, Sphyrna mokarran, Pristis pernotteti, Rhinobatus lentiginosis, Dasyatis guttata, Dasyatis geijskesi, Gymnura micrura, Aetobatus narinari, Rhinotero bonasus, Narcine brasiliensis e Mobula hypostoma.

Esta fauna apresenta afinidades com as associações de espécies do Mioceno no Mediterrâneo.

A area de estudo é um importante criadouro de tubarões da região central do Oceano Atlântico tropical merecendo cuidados especiais e proteção.

This is a survey of elasmobranch fauna native to the Maranhão coast of Brazil (one of the most important breeding grounds for sharks in the Tropical Atlantic). Maranhão's mangrove forests are the tallest in the world and one of Brazil's most structurally complex ecosystems. *Carcharhinus leucas* is one of many shark species that inhabit this area, but Lessa only mentions it in passing. She instead takes on a broader focus by examining the evolutionary significance of the proportionally high numbers of Carcharhinidae in Maranhão's tropical coast system.

The Carcharhinidae dominated Lessa's survey in terms of biodiversity (8 out of 13 sharks and 22 total elasmobranchs) and biomass (77.4% of the total catch), with the Sphyrnidae taking second place on both counts (4 species constituting 18.8% of the catch). This kind of result attributes a more 'modern' character to Maranhão, as the Carcharhinidae and Sphyrnidae are both descendants of evolutionarily recent lineages. With this idea in mind, Lessa turns to a study published by Bassedick et al. in 1984 concerning Miocene sharks of the Mediterranean.

The Miocene environment south of France had a tropical-subtropical climate and vegetation quite similar to that of present-day Maranhão. The sharks associated with this environment were also of a very similar composition to that of modern Maranhão (with the Carcharhinidae, again, being a dominant family over time). As the global climate cooled after the Miocene, the Mediterranean slowly lost its tropical character; consequently, the Carcharhinidae-dominant Miocene sharks of the Mediterranean may have, as Lessa suggests, experienced a 'recent' southwestward range-shift. In this way, these species may have come to Maranhão and other mangrove-lined latitudes to successfully exploit their tropical niche and define their modern elasmobranch identity.

1988

Granulinema gen. n. a New Dracunculoid Genus with Two New Species (G. carcharhini sp. n. and G. simile sp. n.) from the Bull Shark, C. leucas, from Louisiana, USA
Folia Parasitologica, 35 (2): 113-120
MORAVEC, F. & LITTLE, M.D.

Two new species of dracunculoid nematodes, *G. carcharhini* sp. n. and *G. simile* sp. n., representing a new genus *Granulinema* gen. n. are described from the bull shark, *Carcharhinus leucas*, from Louisiana (Lake Borgne), USA; the site of their localization in the host is unknown (probably abdominal cavity). The nematodes of both species were found in tissue juice and only males, juvenile females and body fragments of more advanced but nongravid females were obtained. *Granulinema* gen. n. differs from *Micropleura*, the only other genus in the family Micropleuridae, mainly by the presence of marked, dark excretory corpuscles in lateral excretory canals, pointed tail in females, greater number (6) of postanal pairs of caudal papillae in males, and by the presence of conspicuous transverse cuticular ornamentations on the body surface of mature females. The two new species can be easily distinguished from each other by the length of their spicules; moreover, there are 2 pairs of preanal papillae in the male of *G. carcharhini* sp. n., while there are 3 pairs in *G. simile* sp. n.

Although the bull shark, the host of *Granulinema* spp., of the Atlantic and Pacific oceans is also found in some lakes and streams in Mexico, Central America, and northern South America, occurring e.g. inland up the Amazon River as far as Peru (see Nelson 1976), the *Granulinema* species are apparently marine parasites.

In 1975, thirteen Bull Sharks were caught from Louisiana's Lake Borgne and scanned for parasitic worms (helminths) in an effort to, "obtain little-known tissue-dwelling capillariids reported from requiem sharks as *Capillaria carcharhini* and *C. spinosa*". Amongst the recovered worms were 11 individuals never before known to science. Moravec & Little separated these helminths into two new species of a completely novel genus: *Granulinema carcharhini* (= sharpnose shark granule-nematode) and *Granulinema simile* (= similar granule-nematode).

Like Moravec & Little's original target, the *Granulinema* worms are thought to be tissue parasites. Morphologically, these are Micropleurid worms defined by the presence of dark corpuscles in lateral excretory canals (lending to their genus name) as well as unique secondary sexual characteristics (with males having 6 pairs of postanal papillae and females having a pointed tail and distinct cuticle ornamentation when mature). The two species are distinguishable from one another by the length of their spicules and number of preanal papillae (with *G. carcharhini* having longer spicules and fewer papillae). The worms' family of Micropleuridae is known from only one other genus: *Micropleura*, which is parasitic of crocodiles and turtles. *Micropleura* was not well described by 1988; Moravec & Little's description of *Granulinema* is likewise incomplete, as samples of the gravid female stage could not be found.

Granulinema is a very different type of parasite from the copepods that we have encountered earlier. The worms' small size and internal location makes it quite difficult to tease out their ecological niche. As copepods are conversely large and external, their nature is bit easier to decode (especially when morphological structures such as adhesive-producing maxillipeds clearly reveals a species' natural functions and behavior).

1990

Shark Fisheries in Mexico:
The Case of Yucatan as an Example
In: Pratt, H.L., Gruber, S.H., Taniuchi, T., Editors.
Elasmobranchs as Living Resources: Advances in the
Biology, Ecology, Syst. and the Status of the Fisheries:
NOAA Technical Report NMFS, 90: 427-442
BONFIL, R. & DE ANDA, D. & MENA, R.

Fishing for sharks in Mexico has been a traditional activity for many decades, but little has been done to rationally manage these fisheries. Studies leading to management of the shark fishery of Yucatan, Mexico, were begun in 1984 at the Instituto Nacional de la Pesca. Because no prior information was available on the shark populations under exploitation, the first step has been to determine species composition of the commercial catches, as well as meristics and population structure of the important species. A survey of fishing units used in the local fishery is reported. These can be divided into vessels which fish only for sharks and those which target other species of fishes and land sharks as a bycatch. To date, 25 different species of sharks have been recorded, with five carcharhinids, one sphyrnid, and one triakid being the most important shark species...

Future research should cover life cycles of the main species. Expansion of tagging programs in the area would help define the migratory patterns of the sandbar and dusky sharks, and would help to develop basic information about the migrations of silky and bull sharks, which are poorly understood. This studies would need joint efforts with other scientists and institutions both in Mexico and the United States.

The Bull Shark is one of Yucatan's most commercially important species, composing 27.2% of the total weight of commercial large shark catch and 7.5% of the total number of landed individuals. Bonfil et al. offer the following length-weight equation for this species (sample size: 182; correlation: 0.953):

$$Total\ Weight = 1.1074 \times 10^{-5}\ Total\ Length^{2.9234}$$

The Yucatan population of *Carcharhinus leucas* consists mostly of preadult and adult sharks, but no newborns or early juveniles. Though the largest recorded individual was a female of 265 cm Total Length (TL), most large Bulls in this area are male (with males having a modal class of 235 cm TL and females having a modal class of 195 cm TL). Males are also more numerous than females in Yucatan, with a sex ratio of 1:2.5. According to Bonfil et al., the totality of this information suggests, "that mating and breeding takes place somewhere outside of the fishing grounds of the Yucatan fishery".

Bull Sharks seem to require coastal lagoons for their breeding and nursery areas—a habitat that Yucatan's Bank of Campeche mostly lacks. The authors thus propose that northern Yucatan may be, "only a temporal habitat or feeding ground for this species on their migrations to breeding and mating areas on the nearby coasts of Quintana Roo, Campeche and even Florida." Bonfil et al. go on to name the Laguna de Terminos and the Bahia Chetumal as possible adjacent breeding grounds, and furthermore cite the American Atlantic coast as, "a major breeding and nursery area".

The overharvesting of pups from habitat essential to their growth and development is one of the most important concerns in fisheries science; as such, this information on the Bull Shark's local population structure is critical to the development of a successful and responsible management plan for the species in Mexico.

1990

Fichas FAO de Identificación de Especies para Actividades de Pesca. Guia de Campo das Especies Comercais Marinhas e de Aguas Salobras de Mocambique
FAO, Rome, pp. xxii, 424, 42 pls
FISCHER, W. & SOUSA, I. & SILVA, C. & DE FREITAS, A. &
POUTIERS, J.M. & SCHNEIDER, W. & BORGES, T.C. &
FERAL, J.P. & MASSINGA, A.

Este guia de campo inclui espécies de peixes ósseos, marinhos e de águas salobras, tubarões, raias, lagostas, camarões, caranguejos, estomatópodes, cefalópodes e tartarugas marinhas com interesse commercial em Moçambique. Cada um destes grupos de recursos é precedido de uma introdução sobre a terminologia técnica utilizada, seguida de uma lista de anotações e ilustrações das espécies mais importantes que inclui os nomes FAO, os nomes locais mais vulgares, o tamanho máximo, o habitat e as artes de pesca utilizadas. No sentido de facilitar a primeira identificação, o grupo "Peixes ósseos" é também precedido de um guia ilustrado das diferentes familias.

Carcharhinus leucas (Vale, 1839)   **ver gravura   XL, 259**

Nomes – Nacional: Marracho touro;

FAO: Bull shark

Tamanho: até 350 cm

Habitat e biologia: espécie demersal, lenta, em águas costeiras, baías e estuaries e na parte baixa dos rios; alimenta-se de peixes, pequenos tubarões, raias, crustáceos, ouriços, etc.; muito perigoso; comum

Artes de pesca utilizadas: palangres e redes de emalhe

This FAO fisheries report on Mozambique is just as brief as the previous concerning Angola, but describes *Carcharhinus leucas* a bit differently. The shark is regarded to be a 'very dangerous' demersal species that cruises slowly throughout its coastal habitat. Longline and gillnet fishing are acknowledged as the two most common methods of harvest for Bull Sharks in Mozambique (which is likewise a new development, as the Angolan report only mentioned 'hook and line'). Our first FAO regional term is also provided for this species: "Marracho touro" (Bull Porbeagle).

We need to take a quick step back and reflect: I have recently been reporting on only the raw science of the matter—the facts without the narrative. Science at the close of the 20th Century has fully crystalized into a sharper and clearer institution largely unburdened by the confusions of miscommunication, cultural bias, and prosaic narrative. This is a good place—a waxing of true empiricism; but does it come with a wane?

For example, let us look back in 1859 to Bennett: the scientist who referred to the Bull Shark as a 'monster' within a rather dramatic narrative of a fishermen's duel. His is a science with a tinge of prose—a science that for obvious reasons could not survive in today's stylistic sterilization. I find myself missing Bennett because I was *engaged* with Bennett; his subtle but comparatively rich storytelling triggered an emotional response, namely, sympathy for the dying Bull Shark attempting to fight back against its doom.

I can't foster such an emotional connection with today's scientific format; it is too cold and clean for emotion—too objectively presented to align with any person's sensitivities. This is largely a good thing: facts unbiased and uncorrupted are essential to wise observation, policy, and management. But was the sacrifice for such wisdom an older Romantic merit—a merit that still speaks to us from the grave?

1990

Serum Levels of Circulating Steroid Hormones in Free-Ranging Carcharhinoid Sharks
**In: Pratt, H.L., Gruber, S.H., Taniuchi, T., editors.
Elasmobranchs as Living Resources: Advances in the
Biology, Ecology, Syst. and the Status of the Fisheries:
NOAA Technical Report NMFS, 90: 143-155
RASMUSSEN, L.E.L. & GRUBER, S.H.**

Requiem sharks of the family Carcharhinidae and the closely related sphyrnid or hammerhead sharks are the only lower vertebrates to possess placental viviparity. In contrast to the well-known role of steroid hormones in regulating reproductive processes in mammals, little information exists on the type, amount, seasonal timing, or reproductive variations of steroid hormones in these two shark families. In this study baseline values were measured for four circulating steroid hormones: estradiol, progesterone, testosterone and dihydrotestosterone. Additionally, corticosterone and progesterone, often associated with stress responses, were measured in eleven species of sharks. Of special interest were the elevated serum estradiol and testosterone levels from actively courting female lemon sharks and from female blacknose and hammerhead sharks with large ova in the ovary.

A unique aspect of this paper is that we report for the first time the collection of serial samples from a reproductively active, wild, carcharhinid shark held briefly in captivity. Serum steroid levels from this and one other shark initially fell during confinement.

Blood. Bull Shark blood. Imagine it; a compact, squat, translucent vial gushing with the ultimate symbol of the dichotomy of life and death—a symbol extravagantly amplified by the very form of the shark itself. Like blood, sharks are an ultimate representation of a prime duality; their heterotrophy (poetically described as 'life begat from death') is gloriously apparent with their sleek, graceful, almost angelic forms powered by their perfectly destructive, razor-sharp senses and jaws. Sharks are fueled by blood both literally and symbolically; their life depends on bloodshed and their form is maintained by blood's circulation. Culturally, sharks are almost synonymous with blood—the common trope of a blue ocean stained with red whenever a shark is in the picture. The relationship between sharks and blood is gushing with profundity, making this scientific report by Rasmussen & Gruber even more of a delight to pore.

Within the blood of *Carcharhinus leucas* drift hormones *essential*: 17-β Estradiol (E_2), Progesterone (P), Testosterone (T), Dihydrotestosterone (DHT), and Corticosterone (CS) all dance and mingle to create a profound impact on the shark's courtship, reproduction, development, maturation, ovulation, pregnancy, aggression, and stress.

The Bull's serum steroid hormone concentrations (in pg/mL) are here recorded as:

Stage / Gender	E_2	P	T	DHT	CS
Maturing Female	104	12	121	2	157
Mature Male	9	1,176	2,737	156	1,069
Mature Male	25	4,006	357,540	168,880	3,896

The testosterone reading of 357,540 pg/L for one male Bull Shark is, "among the highest recorded in vertebrate serum". This may suggest an aggression in breeding males.

1990

FAO Species ID Sheets for Fishery Purposes Field Guide to the Commercial Marine Resources of the Gulf of Guinea **Prepared and Published with the Support of the FAO Regional Office for Africa. Rome: FAO. 268 p. SCHNEIDER, W.**

Summary

This field guide includes the marine and brackish-water species of bony fishes, sharks, batoid fishes, lobsters, shrimps, crabs, cephalopods, bivalves, gastropods and sea turtles of present or potential interest to the fisheries of the countries bordering the Gulf of Guinea. Each major resource group is introduced by a general section on technical terms, followed by an annotated and illustrated list of the more important species which includes FAO names, size, fishing gear and habitat. To facilitate the identification, the chapters "Bony fishes", "Sharks" and "Batted fishes" are preceded by an illustrated guide to the different families.

A composite index of scientific and vernacular family and species names is also provided.

Carcharhinus leucas (Valenciennes in Müller & Henle, 1839)
FAO names: **En** – Bull shark; **Fr** – Requin bouledogue; **Sp** – Tiburón sards.
Size: 350 cm, common to 260 cm.
Fishing gear: longlines
Habitat: mainly in coastal waters, also in estuaries and hypersaline waters.
Loc.name(s):

The Gulf of Guinea is pushed and pulled by three major currents: the Benguela, the South Equatorial, and the Guinea. These three primal forces rip the shallow surface waters away from the East African shore, and in so doing summon cooler nutrient-rich water from the deep upwards into the sunlight. This event—known as an upwelling—is a collision of resources and energy; from such a cataclysm, life explodes into unprecedented abundance.

Phytoplankton are the first to bloom. Thriving off of the increased inputs of light and nutrients, they harness the highest amount of energy possible for any organism, and become the foundation for a towering trophic pyramid. Zooplankton swarm swiftly after the phytoplankton and harness their energy into a larger, more accessible package; in response, small fish swoop in to reap the newfound riches. Anchovies, herring, and other planktivores serve as a direct energy link between microscopic and macroscopic life; their massive schools are a compelling beacon to hordes of larger predatory species, and ultimately trigger one of the world's greatest feedings for sharks, seabirds, and man.

Upwelling in the Gulf of Guinea is prominent during the long cold season (June to October) and has a major impact on local artesian fisheries; in 1988, the total marine catch for this region (Côte d'Ivoire to Gabon) was 630,315 metric tons. However, the species composition of this tonnage remains unclear due to the region's high biodiversity, lack of technical training, and common misapplication of species names that make it "practically impossible to estimate the actual fishing effort exercised on most individual species". The amount of Bull Shark harvesting in the Gulf of Guinea is thus unknown. Furthermore, the species' ecological role within the mighty upwelling events of this area is (at this point) still a mystery; does *Carcharhinus leucas* benefit from the bounty?

1991

The Behavior of Sharks in Captivity
Indian Journal of Fisheries, 38 (3): 151-156
HUSSAIN, S.H.

Six species of sharks, viz. *Engomphodus taurus* (sand tiger), *Ginglymostoma cirratum* (nurse shark), *Negoprion brtvirostris* (lemon sharks), *Carcharhinus plumbeus* (sand bar), *C. limbatus* (Pacific black tip), and *C. leucas* (bull shark), were selected to understand some patterns of shark's behavior. The general behavioral patterns like follows, circles and give ways were commonly noted among all the 6 species; however, each species was found to differ from the other. Locomotion in shirk was found to be a combination of anguilliform and crangidform modes. Swimming speed for 6 sharks was measured as [cm/second]. Black tip sharks showed highest speed while sand tigers were found to exhibit lowest speed.

It was significantly observed that behavioural events like follows, circles, hunches and giveways showed marked intraspecific differences. The follows were noted in sand tiger, black tip, sand bars and bull sharks, but not in nurse and lemon sharks. The latter 2 sharks preferred to live alone strolling singly from one end to the other in the tank. Schooling/grouping species followed the rule of going after the more dominant males. Giveway was another important behavioural event but also varied in species. It was observed that dominant shark proceeded straight ahead and the subordinates yielded. The domain sharks among 6 species were bull sharks and sand bars. However, sand tigers did not giveway to these sharks. Hunch and posture in swimming movements were prominent in nurse and sand tiger sharks only.

This incredible study documents five individual Bull Sharks held captive in a 25,000 L tank at Sea World Orlando; the Bulls shared their tank with 27 other individuals (comprising seven different species) and had been observed for over a period of three months. Over that time, it became apparent that *Carcharhinus leucas* is one of the most active species of shark in captivity, with Hussain describing it as, "the most energetic maintaining a uniform speed with no signs of exhaustion."

In the aquarium, Bull Sharks spend most of their time at the surface, occasionally breaking the water with their first dorsal fins; energetic groups were sometimes recorded, "dashing and splashing the water when chasing each other". Patrolling is a very commonly observed behavior, and a larger male usually led the Sea World group of Bulls. This may be an exemplification of dominance, which is both intraspecific and interspecific for *Carcharhinus leucas*. In cases of 'face-to-face' encounters (i.e., when the swimming path of two sharks intersects, forcing them to approach each other and eventually 'meet'), larger Bull Sharks would force smaller ones to yield. Similarly, Bull Sharks would force other species to change their course and adopt recessive behavior.

From these observations we gain a magnetically charged behavioral portrait of *Carcharhinus leucas*. The shark comes off as a living dynamo—a non-stop powerhouse ignorant of obstacle, whose energy induces compelling changes throughout the system (i.e., the forcing of more docile or inactive species to yield or avoid the Bull Shark altogether). This dominance-based behavior is curiously relatable to the human narrative; we can strangely identify the Bull as a 'bully', given its uncompromising attitude and dominance-based social hierarchy. This familiarity is a bit hilarious, but mostly touching; a bridge between our two species' psychologies has now been retrieved from the dark.

1993

Comparisons of Shark Catch Rates on Longlines Using
Rope/Steel (Yankee) and Monofilament Gangions
Marine Fisheries Review, 55 (3): 4-9
BRANSTETTER, S. & MUSICK, J.A.

During the months of June through September in 1991 and 1992,
71 shark longlines were fished in the Chesapeake Bight region of the U.S.
mid-Atlantic coast with a combination of rope/steel (Yankee) and
monofilament gangions. A total of 288 sharks were taken on 3,666
monofilament gangions, and 352 sharks were caught on 6,975 Yankee
gangions. Catch rates between gear types differed by depth strata, by
month, and by species. Analyses were divided between efforts in the
nursery ground of the sandbar shark, *C. plumbeus*, in Chesapeake Bay and
efforts outside the Bay. Mean catch per unit effort (CPUE) ± SE, as sharks
caught per 100 hooks fished, was significantly (P,0.05) lower for Yankee
gangions. Mean CPUE's for sandbar sharks in the nursery ground were
20.6 ± 3.8 for Yankee gangions and 26.0 ± 3.0 for monofilament gangions,
and mean CPUE's for all species combined outside the Bay were 3.7 ± 0.7
for Yankee gangions, and 6.9 ± 1.2 for monofilament gangions.

In addition to monitoring the status of the shark populations of the
Chesapeake Bight, the VIMS program has been a source of specimens for
related research projects. During the course of the program, the catch per
unit effort (CPUE) has declined approximately 75% (Musick et al., 1993).
The large-coastal shark stock of the northwest Atlantic has been overfished
for almost a decade (NMFS, 1993), and the declining CPUE of the VIMS
survey can be at least partially attributed to this.

This Virginia Institute of Marine Science (VIMS) study found monofilament gangions to be ultimately more efficient at catching sharks than the standard 'Yankee' steel gangions (which composed the original longline rigging for the VIMS shark survey). For both the Chesapeake Bay and Virginia Coast, Monofilament outperformed steel in CPUE (Catch Per Unit Effort), here measured as:

$$\frac{Number\ of\ Sharks\ Caught}{100\ Hooks}$$

In regards to the specific study areas, monofilament claimed a significant lead in the Lower Chesapeake Bay, Nearshore (10-20m), and Mid-Shelf (20-100m) with a suggested lead in the <10m Coastal area. Steel outcompeted monofilament in only the Offshore (>100m) study area (though prior research conducted by Berkeley & Campos in 1988 found monofilament to be the more efficient offshore fishing choice).

Only one Bull Shark was captured in this VIMS survey (Lower Chesapeake Bay, monofilament). This is most likely due to the species' limited (temperature-restricted) distribution in the cooler waters of the Virginian Atlantic. Conversely, the closely related Sandbar Shark (*Carcharhinus plumbeus*) abounds in the Virginian, and uses the Chesapeake Bay as a major pupping ground. Within the VIMS survey, Sandbars emerged as the most abundant species overall (composing 55% of the total catch). Sandbar Sharks also completely dominated the Chesapeake Bay catch in terms of total number of individuals captured (160 via steel, 107 via monofilament). Only a single specimen each of *Carcharhinus leucas*, *Odontaspis taurus*, and *Sphyrna lewini* were otherwise recorded in the Chesapeake. Outside the Bay, the Atlantic Sharpnose Shark (*Rhizoprionodon terraenovae*) dominated the Virginia Coastal catch (104 via steel, 93 via monofilament) with Sandbars being the second most commonly acquired coastal species.

1993

Trends in Shark Abundance from 1974 to 1991 for the
Chesapeake Bight Region of the US Mid-Atlantic Coast
In: Conservation Biology of Elasmobranchs. Ed. By S.
Branstetter. NOAA Tech. Report NMFS, 115: 1-18, figs 1-9
MUSICK, J.A. & BRANSTETTER, S. & COLVOCORESSES, J.A.

Recent stock assessments indicate that the shark stock of the western North Atlantic is exploited at a rate twice the maximum sustainable yield. This finding is supported by data generated by the Virginia Institute of Marine Science longline program for sharks of the Chesapeake Bay and adjacent coastal waters. Trends in catch per unit of effort since 1974 indicate 60-80% reductions in population size for the common species — sandbar (*Carcharhinus plumbeus*), dusky (*C. obscurus*), sand tiger (*Odontaspis taurus*), and tiger (*Galeocerdo cuvier*) sharks. Declines include numbers of individuals for all species, size classes within species, and in one case a strong decline in relative abundance. Given the limited ability of sharks to increase their population size, these results suggest that stock recovery will probably require decades.

In the recent past sharks were underutilized; 58% of the estimated recreational and commercial catch was discarded (Hoff and Musick, 1990). Apparently, however, they were not underexploited. Since 1980, the combined recreational and commercial fishing mortality has averaged 22,000 t/year (NMFS[2]); however, MSY for U.S. waters was estimated at 9,800-16,250 t (Anderson, 1985; Parrack, 1990), therefore mortality was 1.5-2.0 times MSY. This over-exploitation is reflected in the declining CPUE for both juveniles and adults of the primary species taken in the Chesapeake Bight region of the mid-Atlantic coast.

94

I am responsible for the shark fisheries problem. I am a heterotroph: by my nature, I consume others to survive, be they plant or animal. I can process many things and convert them into my body—chickens, oranges, grains, scallops, sharks—I break these down to build myself. I am a natural animal; I am allowed to do this.

I'm not particularly good at acquiring fish, but I like the taste of seafood, and am therefore willing to pay others to acquire my fish. I do this at the restaurant, at the supermarket, or even at the dock; I give them my money, and they go out to do my task. They do it because I pay—because I provide. They, too, need to provide.

I am one of many. I am one of millions who do the same as I. Together we have created The Fishermen, the hooks, the nets, the boats, the industry, the debasement, the regulations, the conservation goals, and the Fisheries Scientist. This is our system.

We are the market. We are the foundations of all argument and demonization. The Fisherman and the Fisheries Scientist have sometimes come to blows because of us; their compromises and the very need to make them are direct descendants of our appetite. The fishing culture, economy, and livelihood is of our design; so too is the dying sea.

Every step we take economically is a bite out of the environment. Every step we take environmentally is a blow to economics. These are restless spirits in constant contention, and their balance is maintained by one primal force: our choice.

If I choose to eat fish—to simply select a single sea-item on a restaurant menu—I fund the hooks that kill the sharks. If I choose to avoid fish—to remove the incentive and thereby protect the ocean from my appetite—I depreciate The Fisherman: an important cultural and economic pillar to our civilization. *I am the force that guides each fate.*

1994

Identifying Isolated Shark Teeth of the Genus
Carcharhinus to Species: Relevance for Tracking Phyletic
Change Through the Fossil Record
American Museum Novitates, 3109: 1-53, 18 fig., 1 tabl.
NAYLOR, G.J.P. & MARCUS, L.F.

In this study we investigated the extent to which the isolated teeth
of extant *Carcharhinus* can be correctly assigned to species using
discriminant function analysis of linear measurements. We measured
12,647 extant teeth representing shape variation due to species, ontogeny,
sexual dimorphism, jaw, and tooth position. Observations were split into a
"training" data set used to establish the discriminant functions and a "test"
data set used to evaluate their efficiency. We found that excellent
discrimination could be achieved for teeth from certain parts of the
dentition. Results indicate that teeth from the upper jaw are correctly
assigned to species more often than are teeth from the lower jaw and that
teeth from central positions within a tooth series (half-way between the
symphysis and the angle of the jaw) are correctly assigned to species more
often than are teeth from other positions. Quadratic discriminant analysis
was used to assess whether or not the jaw and tooth positions of isolated
Carcharhinus teeth could be determined. While classification accuracy
varied across species, results indicate that upper jaw teeth can be readily
distinguished from lower jaw teeth and that "maximally distinctive teeth"
can be distinguished from teeth at less distinctive positions within the jaw.
These findings are used to propose an objective and quantitative protocol
for the interpretation of phyletic change in fossil *Carcharhinus* teeth.

Pouring through this incredibly complex dataset, I found myself a bit disappointed to find out that no specific evolutionary relationships between the various species of *Carcharhinus* were, in fact, proposed; instead, only the methodology *used* to describe such relationships was proposed (and in excruciating detail, which I will not replicate here). Essentially, thousands of teeth taken from 30 extant species of *Carcharhinus* were measured in almost every way possible; the chief results of such measurements were: A; teeth in 'the middle' of the upper jaw were the most effective in distinguishing species from one another, and B; such teeth are the ideal candidates for deciphering the rich *Carcharhinus* fossil record (and subsequently piecing together the genus phylogeny).

Though Naylor's report is only a key to the treasured *Carcharhinus* phylogenetic tree, it nonetheless teases some important relationships between *C. leucas* and its requiem shark cousins. Firstly, *C. leucas* teeth were only confused with those of *C. amboinensis* (Pigeye Shark), *C. galapagensis* (Galapagos Shark), *C. longimanus* (Oceanic Whitetip Shark), and *C. brachyurus* (Bronze Whaler), with *C. amboinensis* being the most common case of mistaken identity. Could these observations be reflective of a closer evolutionary kinship (especially in consideration of the famously similar physical appearance of *C. amboinensis* to *Carcharhinus leucas*)?

Of particular interest is the comparative size of Bull Shark teeth to other *Carcharhinus*: they are gigantic. Adult *Carcharhinus leucas* teeth are only rivaled in proportional size by *C. longimanus* and possibly *C. plumbeus*. What this means is that, within *Carcharhinus*, the Bull Shark demonstrates a frightening functional *exclusivity*; only Bulls, Oceanic Whitetips and Sandbars wield the jaws designed to dismember larger prey. Eerily enough, two of these sharks belong to the rare category of 'man-eater'.

1995

Chondrichthyes and Osteichthyes from the Early Pleistocene Leisey Shell Pit Local Fauna, Hillsborough County, Florida

Bulletin of the Florida Museum of Natural History, 37 Pt. 1 (8): 251-272

SCUDDER, S.J. & SIMONS, E.H. & MORGAN, G.S.

The early Pleistocene (early Irvingtonian) Leisey Shell Pit ichthyofauna was recovered from two large commercial shell pits located less than 1 kilometer inland from Tampa Bay in Hillsborough County, Florida. The combined fish fauna from the two Leisey sites is composed of 73 species, including 23 species of sharks and rays and 50 species of bony fish. This is the largest fish fauna ever reported from the Cenozoic of Florida, and includes 34 taxa that represent new additions to the fossil record of the state. There are four extinct taxa in the fauna, all of which are Chondrichthyes: the mako shark *Isurus hastalis*, the nurse shark *Ginglymostoma serra*, the snaggletooth shark *Hemipristis serra*, and the guitarfish *Rhynchobatus* sp. The genera *Hemipristis* and *Rhynchobatus* are now restricted to the Indo-West Pacific region. The two sites comprising the Leisey Shell Pit Local Fauna, Leisey 1A and Leisey 3A, have somewhat different faunas. The most common fish from Leisey 1A in decreasing order of abundance are: alligator gar *Atractosteus spatula*; snook *Centropomus* sp.; mullet *Mugil* sp.; bull shark *Carcharhinus leucas*, and eagle ray *Myliobatis* sp. These species, as well as the majority of the remaining fauna, suggest a shallow marine or estuarine environment such as a coastal bay or mouth of a large river.

One million years ago, there was *Carcharhinus leucas*. It was a prehistoric species – an animal well in existence before the human narrative—and it has survived into our modern Anthropocene via an unbroken bloodline. To imagine its presence in the Leisey site (then a subtropical, shallow estuary habitat) is to imagine a sort of 'virgin' world. Though planetary immaculacy is, in truth, an illusion (as we, like all animals, make natural alterations; buildings, plastic, fast food, all seemingly 'removed' things, are merely complex refinements of our environment), there is a real matter to the idea of human imbalance. Every progression of the human race is a detraction of the other 'races', and it from this fact that the concept of pre-human 'virginity' rings true.

To imagine that a species has been in existence well before our own time is to attribute a romantic 'knowledge' to that species; *C. leucas* can serve as a strange bridge between the pre-human and post-human planet, and its quality of bearing witness to both conditions may be appropriated to an unfamiliar definition of wisdom. No individual Bull Shark has, of course, seen both worlds...but the idea that *C. leucas* as a concept has influenced both the Virgin Planet and the Human Planet is an eloquent testament to the species' *understanding* of our dynamic Earth.

The identity of *Carcharhinus leucas* is an accumulation of a million years of change, and yet, biologically stable: though the Leisey 1A Shell Pit site is 1.0-1.5 million years old, 53.4% of the fossilized Chondrichthyes were still recognizably *Carcharhinus*, with, "the great majority of the *Carcharhinus* teeth present (>90%)" belonging to *Carcharhinus leucas*. This species integrity is indicative of an acute reading of Earth's more balanced roles— niches that change gently in the curl of revolutions. Thus, with such an ear for the planet's tunes, *Carcharhinus leucas* succeeds as a rhyme with time.

1996

Les Peuplements de Poissons des Milieux Estuariens de l'Afrique de l'Ouest: L'Exemple de l'Estuaire Hyperhalin du Sine-Saloum

Université de Montpellier II. Thèses et Documents Microfiches No. 156. ORSTOM, Paris. 267 p.

DIOUF, P.S.

Résumé

Le déficit pluviométrigue qui sévit dans le Sahel depuis plusieurs décennies, combiné à une forte évaporation et à une faible pente de la partie aval de l'estuaire du Sine-Saloum (Sénégal) a été à l'origine de l'inversion du gradient de salinité et de son corollaire l'hyperhalinité.

Deux questions essentielles se posent. Ces modifications de l'environment, et en particulier la sursalure, ont-elles eu un effet negative sur les ressources halieutiques ? Ont-elles modifié les fonctions écologiques fondamentales de l'estuaire: nourricerie, zone de reproduction pour certaines espèces, enrichissement du milieu côtier ?

Une approche comparative avec d'autres milieu estuariens et lagunaires de l'Afrique de l'Ouest a permis de constater que malgré l'hyperhalinité, les biomasses et la richesse spécifique des peuplements de poisons de l'estuaire du Sine-Saloum sont relativement élevées.

Par ailleurs, il est apparu que toutes les fonctions écologiques de l'estuaire sont encore convenablement assures, à l'exception de l'enrichissement du milieu côtier adjacent, à cause de l'absence de crue et d'une forte diminution des apports d'eau douce.

Sine-Saloum is a Senegalese estuary that lies to the west of the semiarid sands of Sahel. Its usual character is like that of all estuaries (saltier water towards the mouth, fresher water towards the source), but two key attributes can completely deconstruct such a familiar regime: a distinctly weak watershed slope and an intense rate of evaporation. When these combine with the effects of Sahelian drought, the salinity gradient of the Sine-Saloum completely inverts: fresh water burns away to leave behind a higher concentration of salt, causing the inner portions of the estuary to become 'hyperhaline' (greater than 40ppt in salinity, which is higher than ocean water).

As the salinity paradoxically *increases* upriver, the biodiversity and biomass of inland fauna decreases, leaving only a simple trophic structure of mostly scavengers upstream. However, the resulting lack of freshwater (which usually dilutes the Sine-Saloum estuary) creates a positive trophic benefit downstream: nutrients that would have otherwise been 'flushed out' via freshwater flooding are instead retained at the estuary mouth, creating an energy-rich environment for marine species capable of handling higher salinities. Groupers, mackerel, mullet and others directly benefit from this resource retention, and thereby increase in their abundance; ultimately, their growth in numbers offsets the upstream reduction of faunal diversity and abundance (thus preserving the net biodiversity and biomass of the Sine-Saloum estuary as a whole).

Diouf does not specifically describe how the Bull Shark plays into this unusual dynamism. However, the Sine Saloum's hyperhalinity creates a unique opportunity for this shark; as upriver fish cannot tolerate the drastic increase in salinity, they yield their ecological niches to more elastic species. Bull Sharks—with their euryhaline tolerance of up to 50ppt—can easily exploit this advantage (and perhaps have already done so).

1996

Fishes of the Cambodian Mekong
FAO Species ID Field Guide for Fishery Purposes
FAO, Rome, 265 p.
RAINBOTH, W.J.

This field guide includes approximately 500 species of present or potential interest to, or likely to be encountered in, fisheries in the Cambodian Mekong. This covers all species historically recorded from Cambodian reaches of the Mekong as well as numerous new records by the author. It also includes species expected to occur in Cambodia which have been recorded from, or found by the author, in Vietnam, Thailand and Laos. An overview is given on the factors that have contributed to the diversity of the Mekong along with a basic introduction to taxonomy. A section on technical terms and measurements illustrates the characters used for the identification. As an aid to identification to higher taxonomic levels, an illustrated guide to orders and families is included. The species accounts provide information on synonyms and misidentifications, FAO (English) names, Cambodian names (in Cambodian script), sizes, diagnostic features, one or more illustrations, and notes on fisheries, distribution, habitat and biology. The guide is fully indexed and a list of further literature is appended. Finally, 216 colour photographs of fishes of the Cambodian Mekong are presented.

Carcharhinus leucas (Valenciennes, 1839)
Local names: Trey chhlarm
Size: To 100 cm in fresh water, up to 300 cm in the sea.

The story of Mekong begins with the story of the Sunda Shelf; an extended continental platform that was most likely exposed during the glacial ages of the Pleistocene. Old rivers carved both valleys and unique species assemblages into the ancient Sunda, creating distinct, isolated aquatic communities. Water, however, chose to rise; escaping its glacial confines, the sea expanded and swallowed much of the Sunda Shelf to separate Southeast Asia from the modern islands of Sunda, Java, Sumatra, and Borneo. With this massive change came unification; the once separate river valleys drowned, and yielded a new, broader, more mixed marine environment in which once-distinct aquatic communities mingled, adapted, and evolved in coexistence.

Fire poured across southeastern Cambodia 600,000 years ago; lava flows altered the landscape as consequence of a fan-shaped fault system—a system that possibly encouraged the continental subsidence of southern Cambodia and Vietnam. Earth's volatile tectonic activity forced multiple streams to collide and form a larger, unprecedented entity: a river that by 10,000 years ago, poured quickly into the South China Sea, and by 5,700 years ago, bore a Great Lake, a rich estuary, and one of the most biodiverse communities on the planet. This is origin of Mekong: child of the elements.

The life of Mekong is in direct harmony with its history, as "the extensive re-alignment of drainage configurations, much of which has occurred very recently, geologically speaking, has mixed formerly isolated species assemblages repeatedly." As a result of this unique incorporation of "diversity that has evolved in distant areas", an estimated 1,200 fish species call home to this river. *Carcharhinus leucas* has not yet been officially recorded, but Mekong's richness and mystery compels the biologist (and the FAO) to easily imagine the Lord of Estuaries as an integral piece of the river's path.

1996

FAO Species ID Field Guide for Fishery Purposes
The Living Marine Resources of Somalia
FAO, Rome. 376 p.
SOMMER, C. & SCHNEIDER, W. & POUTIERS, J. M.

This field guide covers the major resource groups likely to be encountered in the fisheries of Somalia. This includes shrimps, lobsters, bivalves, gastropods, cephalopods, sharks, batoid fishes, bony fishes, and sea turtles. Each resource group is introduced by a general section on technical terms and measurements pertinent to that group and an illustrated guide to orders and families of the group. The more important species are treated in a subsequent guide that includes scientific nomenclature, FAO names in English and French (were available), local names used in Somali, diagnostic features, one or more ullustrations, maximum suze, and notes on fisheries, habitat, and biology. The guide is fully indexed and a list of further literature is appended.

Carcharhinus leucas (Valenciennes in Müller & Henle, 1839)
FAO names: En – Bull shark; **Fr** – Requin bouledogue.
Size: To 3.5 m.
Fisheries: Caught with longlines and gillnets.
Habitat and biology: In coastal, estuarine and lacustrine waters, usually found close inshore in marine habitats; occurs from depths of 1 to 152 m. Omnivorous, feeding mainly on fishes, invertebrates, and carrion.
Remarks: Known to be dangerous to people, probably one of the most dangerous sharks.

The character of Somali marine life is largely locked away from empirical eyes due to region's oceanographic and political instability. The ongoing Somali Civil War was a decade old by the time of this FAO publication, and has compromised the ability to conduct ichthyological fieldwork (as well as the accessibility of vital information such as precise assessments of the region's annual fisheries production). It is estimated that the total annual catch of Somalia never exceeded 20,000t, though the potential annual catches could be 200,000t to 300,000t. Indeed, by the 1980s the Somali government was keen to further develop fisheries, but the civil war interrupted this progression, and rendered most of the fishing gear, processing plants, and equipment, "either lost or destroyed".

Divisiveness and tumult in Somalia are not exclusive to the social realm. Ecologically, the region is highly variable in terms of its currents, salinity, and temperature, and is furthermore split into two separate tropical zones (the Red Sea and Arabian Peninsula to the north, and the east African tropical coast to the south) by a seasonal cold-water upwelling area at the center. The Somali Current (the world's fastest, with speeds of up to 7 kn) pushes northeastward and periodically creates in its wake a magnificent gyre known as the "Great Whirl"; it is this phenomenon that subsequently induces the powerful upwelling that divide the North and the South.

The volatility of Somalia, the elementality of Mekong, and the hyperhalinity of Sine-Saloum can all be tied together by the existence of *Carcharhinus leucas*. Like every organism, the Bull Shark derives its form from the tutelage of its environment; with seascapes as harsh and powerful as the Somali, Cambodian, and Senegalese coasts, it is no small wonder that our shark has likewise evolved into a robust and powerful species. What has allowed *Carcharhinus leucas* to master so many challenging disciplines?

1997

The Living Marine Resources of Kuwait, Eastern Saudi Arabia, Bahrain, Qatar, and the United Arab Emirates
FAO, Rome
CARPENTER, K.E. & KRUPP, F. & JONES, D.A. & ZAJONZ, U.

This field guide covers the major resource groups likely to be encountered in the fisheries of Kuwait, Eastern Saudi Arabia, Bahrain, Qatar, and the United Arab Emirates. It includes marine plants, shrimps, lobsters, crabs, bivalves, gastropods, cephalopods, sharks, batoid fishes, bony fishes, sea snakes, sea turtles, sea birds, and marine mammals. In order to serve as a tool for ecological and biodiversity studies, all species known from the Gulf of certain groups are included. These include the sharks, batoid fishes, bony fishes, sea turtles, and marine mammals. Each resource groups is introduced by a general section on technical terms and measurements pertinent to that group and an illustrated guide to hgher taxonomic groups when relevant. Species are then treated in a subsequent guide that includes scientific nomenclature common English and Arabic names where available, size information, information on habitat, biology, and fisheries, diagnostic features, and one or more illustrations, some of which are included in colour. The guide is fully indexed and a list of references is appended.

Carcharhinus leucas (Valenciennes, 1839)
Synonyms/misidentifications: *Carcharhinus zambezensis* (Peters, 1852).
Common names: **En** – Bull shark; **Ar** – Jarjur
Size: Maximum total length 340 cm.

106

The Persian Gulf is a perfect subject for the continuation of our last discussion. *Carcharhinus leucas* is one of a few select species that can tolerate the Gulf's many bizarre and challenging elements. It is a place of environmental extremes and riches—a young, volatile seascape with rewards fit only for the capable few. 20,000 years ago, it was nothing more than an arid basin. Only with the modern rise in sea level did the Gulf begin to take its present shape as a habitat of contrasts—a confusingly paradoxical seascape that both poisons and presents new opportunities for life.

The current marine denizens of the Persian Gulf can all trace their ancestry back to a modern Indo-Pacific recolonization event. This movement was drastically limited by the Gulf's geography, as the Straight of Hormuz – the 'entrance' to the Gulf—would select against the dispersal of marine larva from adjacent seas. Bone chilling winters and blistering summers from the basin's surrounding lands create a massive water temperature range of 29°C. With such extremes, evaporation becomes a major concern, as escaping seawater leaves its salt behind to contribute to astronomical salinities (50 to 70ppt on average, 200ppt maximum) within the enclosed coastal areas.

A final factor in shaping the character of the Persian Gulf fauna is oil; vast fields beneath the seabed naturally seep their contents into the water column. Though many organisms respond negatively to this chemical influx, some (such as bacteria) are grateful, and convert the oil into a source of organic carbon. This resource— in conjunction with the ubiquitous amount of sunlight and massive nutrient mixing via prevailing seasonal winds—owes to the Persian Gulf's uncanny character of high productivity. It is one of the richest ecosystems in the world for marine benthos, and boasts an incredibly developed fisheries market for shrimp, pomfret, shad, and grunter.

1997

Catch and Bycatch in the Shark Drift Gillnet Fishery off
Georgia and East Florida
Marine Fisheries Review, 59 (1): 19-28
TRENT, L. & PARSHLEY, D.E. & CARLSON, J.K.

An observer program of the shark drift gillnet fishery off the
Atlantic coast of Florida and Georgia was begun in 1993 to define the
fishery and estimate bycatch including bottlenose dolphin, *Tursiops
truncatus*, and sea turtles. Boats in the fishery were 12.2-19.8 m long. Nets
used were 275-1,800 m long and 3.2 – 4.1 m deep. Stretched-mesh sizes
used were 12.7 – 29.9 cm. Fishing trips were usually <18 h and occurred
within 30 n.mi. of port. Fishing with an observer aboard occurred between
Savannah, Ga., and Jacksonville, Fla., and off Cape Canaveral, Fla. Nets
were set at least 3 n.mi. offshore. Numbers of boats in the fishery increased
from 5 in 1993 to 11 in 1995, but total trips decreased from 185 in 1994 to
149 in 1995. During 1993-95, 48 observer trips were completed and 52 net
sets were observed. No marine mammals were caught and two loggerhead
turtles, *Caretta caretta*, were caught and released alive. A total of 9,270
animals (12 shark, 21 teleost, 4 ray, and 1 sea turtle species) were captured.
Blacknose, *Carcharhinus acronotus*; Atlantic sharpnose, *Rhizoprionodon
terraenovae*; and blacktip shark, *C. limbatus*), were the dominant sharks
caught. King mackerel, *Scomberomorus cavalla*; little tunny, *Euthynnus
alleteratus*; and cownose ray, *Rhinoptera bonasus*, were the dominant
bycatch species. About 8.4% of the total catch was bycatch. Of the totals,
9.4% of the sharks and 37.3% of the bycatch were discarded.

'Harvest' is a nourishing word; it appeals to both the soul and the stomach as a symbol of bounty, health, and security. The cornerstones of wellbeing are built upon the foundations of 'harvest'; it is why autumn is the season of good fortune and feasts. It is thus natural to feel a strangely positive notion in regards to the Harvest of Sharks. We (and the fishermen that we employ) are each engaged in a powerful act: the drawing of bounty from the greatest well—the tapping of seas for our own sustainment. This is no small accomplishment, but a mighty predation; upon the complex and challenging platform of the ocean we have constructed a titanic identity. Our power derives, in part, from sharks. We are their predators, and we accomplish our harvest via the following key techniques:

Pelagic Longline	*Hooks suspended on snoods from a mainline: surface*
Bottom Longline	*Hooks suspended on snoods from a mainline: seabed*
Gillnet	*Nets shaped like a panel and towed in a straight line*
Rod and Reel	*Hooks suspended from a single line on a single rod*
Handline	*Hooks suspended from a single line (no rod)*
Bandit Gear	*Hooks suspended from a single line on a reel*

Unfortunately, this is not a victimless power. Sharks are alive: they feel the pain of dismemberment, entanglement, and struggle—the horrors of defeat. They are essential components of a dying ecosystem, and to overharvest sharks is to weaken our own foundation—to poison our own sense of bounty, health, and security. Without sharks, the platform would crumble, and our power, wellbeing, stomach, and souls would, likewise, collapse. This report is concerned primarily with the bycatch of marine mammals and sea turtles, which are each heavily protected: will sharks someday bear the same concern?

1998

FAO Species Identification Guide for Fishery Purposes
The Living Marine Resources of the W. Central Pacific
Vol. 2 Cephalopods, Crustaceans, Holothurians and Sharks
Rome, FAO. 1998: 687-1396
CARPENTER, K.E. & NIEM, V.H.

Carcharhinus leucas (Valenciennes <u>in</u> Müller and Henle, 1839)

Habitat, biology, and fisheries: Predominantly a coastal and fresh-water species inhabiting shallow waters, especially in bays, river estuaries, rivers, and lakes. It tolerates a wide range of salinities, readily penetrates far up rivers and also into hypersaline bays. Usually slow-swimming if active while cruising, this bottom-living shark may develop great speed when chasing its prey. Viviparous, number of embryos up to 12. The young readily tolerate low salinities, and some are born in fresh water. An opportunistic predator with a very wide food spectrum that includes bony fishes, sharks, rays, invertebrates (crabs, shrimps, sea urchins, etc.), marne and freshwater turtles, birds, marine and terrestrial mammals, and carrion. It has large strong jaws and large stout teeth for its size, which enable it to dismember and feed on relatively large prey. Known to be dangerous to people, and possibly one of the most dangerous sharks because of its inshore and fresh-water habitat, large size, powerful feeding structures, and omnivorous habits. Caught mainly with longlines and gill nets and used for its meat, hide, fins, liver oil, and as fishmeal.

Distribution: Widespread along the continental coasts of all tropical and subtropical seas; also, the most wide-ranging cartilaginous fish in fresh water.

Carpenter & Niem's FAO Species Identification Guide is considerably more detailed than the previous FAO publications in regards to diagnostics, habitat, biology, and fisheries importance. Each species profile is a precisely balanced formula of breadth and depth, and for *Carcharhinus leucas*, only the essential marks are noted. This succinct and sufficient portrait—a product of over a century's worth of information refinement—is a trim and clean approach to the 'modern Bull Shark': the basic foundation of today's common understanding of the species. Physically, this shark's principles are defined as:

- "Snout very broadly rounded and extremely short"
- "Teeth in upper jaw triangular, with broad, heavy, serrated cusps"
- "First dorsal fin high and broad with a pointed or slightly rounded apex, its origin a little in advance of insertion of pectoral fins; second dorsal fin high with a short posterior lobe"
- "Pectoral fins broad, with narrow pointed tips. No interdorsal ridge"

FAO is an excellent reference for standardization, and I wager that their interpretation of 'essential characters' is the most precise. I understand that you may be jaded by the constant repetition of these fundamentals, but such drilling is essential for comprehending the *definition* of this species. What we are tying to achieve—and what we have always been achieving—is a grasp of *uniqueness* and *identity*. Without each, there would be no Bull Shark; there would be no special, irreplaceable combination of characters, no new flavor to the universe—no speciality. Speciality is enrichment; the world is exciting only through the images, voices, challenges, sensations, and experiences of speciality. As each elemental contributes to the glittering fabric of the universe, we must celebrate *definition* and *identity* in every form—in *Carcharhinus leucas*, whose voice is distinct in the void.

1998

NMFS Cooperative Shark Tagging Program, 1962-93: an Atlas of Shark Tag and Recapture Data-NMFS **Marine Fisheries Review, 60 (2): 1-87** **KOHLER, N.E. & CASEY, J.G. & TURNER, P.A.**

The National Marine Fisheries Service (NMFS) Cooperative Shark Tagging Program (CSTP) is part of continuing research directed to the study of the biology of large Atlantic sharks. The CSTP was initiated in 1962 at the Sandy Hook Laboratory in New Jersey under the Department of Interior's U.S. Fish and Willife Service (USFWS). During the late 50's and early 60's, sharks were considered a liability to the economy of resort communities, of little or no commercial value, and a detriment to fishermen in areas where sharks might damage expensive fishing gear or reduce catches of more commercially valuable species.

Several shark attacks along the New Jersey coast at that time gave rise to public concern about a perceived shark menace. In response to that concern, a shark longline survey was conducted in 1961 from Jones Inlet, N.Y., to Cape Henlopen, Del., by laboratory staff. The objectives of that study were to determine the species composition, distribution, abundance, food habits, seasonal occurrence, and other aspects of the biology of large sharks off the middle Atlantic states. The survey resulted in the capture of over 300 sharks, including white sharks, *Carcharodon Carcharias*; and tiger sharks, *Galeocerdo cuvier*, considered to be among the most dangerous species.

This paper broadly summarizes the tagging and recapture (T/R) information from the CSTP for 1962 through 1993.

American tagging studies usually display an interesting trend of Bull Shark infrequency, with Bulls often making up only <1% of the total catch (as in this NMFS Coastal Shark Tagging Program report). According to the Kohler, Casey, & Turner, "lower tagging and recapture success does not necessarily reflect low abundance but may mean that a species may be undesirable or inaccessible to the main body of fishing and tagging effort." Though I wouldn't necessarily call the Bull Shark a 'desirable species', it is nevertheless an NMFS-permitted large coastal shark with a unique preference for the habitats closest to human activity and influence: given its availability and proximity, I find the ideas of either undesirability or inaccessibility surprising. Could the Bull then perhaps be a truly uncommon species? Or maybe, is it shark with an oddly selective preference against longlines, gillnets, rod and reel, or other common fishing methods? Does this species naturally shy away from anglers more often than others sharks?

Whatever the explanation for this relative infrequency, *Carcharhinus leucas* still has a significant presence within the NMFS CSTP database; 520 Bull Sharks (177 male, 249 female, 94 unknown gender) were tagged over the 32-year period, with only 10 being recaptured (recapture rate of 1.9%). A maximum speed of 1.6 nautical miles per day was recorded for this species, along with a maximum distance traveled of 235 nautical miles.

Like many sharks in this survey, *Carcharhinus leucas* was also noted to leave the 200-mile American Exclusive Economic Zone (EEZ). This documentation is critical for the development of proper management of 'transboundary' shark stocks. In this survey alone, "fishermen representing 32 countries have tagged sharks and 47 countries are represented in the tag returns." Like many American Atlantic sharks, Bulls are subject to multinational fisheries exploitation, and must be managed with international cooperation.

1998

The Bull Shark, Carcharhinus leucas (Valenciennes, 1841),
from the Usumacinta River, Tabasco, Mexico, with Notes
on its Serum Composition and Osmolarity
Ciencias Marinas, 24 (2): 183-192
SOSA-NISHIZAKI, O. & TANIUCH, T.
& ISHIHARA, H. & SHIMIZU, M.

Five specimens of the bull shark, *Carcharhinus leucas*, were collected from a freshwater lagoon (2-3 ppt) and the Usumacinta River in Tabasco, Mexico, in May 1993. The serum osmoregulatory status of two individuals was analyzed: a mature 14-year-old male (2180 mm total length, the largest bull shark reported in fresh water) and a neonate (786 mm total length). Blood serum ionic concentrations of sodium chloride and calcium were reduced and compared with the seawater and marine forms. Urea serum contents were significantly lower than those reported for other bull sharks living in marine and freshwater environments, but much higher that those obtained for true freshwater elasmobranchs, such as potamotrygonid stingrays (Potamotrygoninae). Osmolarity was approximately 70% that of seawater. This osmoregulatory status suggested that the large bull shark analyzed had a longer time of residence in fresh water than other specimens reported in the literature.

This is the first documented description of the occurrence of the bull shark in the Usumacinta basin, Tabasco, Mexico. Castro-Aguirre (1978) mentioned that the coastal area around Frontera, Tabasco, was one of the localities of the bull shark.

The Bull Shark's power to enter and exploit freshwater habitat is a complicated mechanism of uncertain origin. Perhaps the strongest theory as to 'why' *Carcharhinus leucas* has evolved with such a power is that, "the bull shark takes advantage of an ecological opportunity, pursuing food inside any available freshwater channel, where there is almost no competition". Explanations as to 'how' can be difficult to parse, but important clues can be coaxed from the composure of blood.

Sosa-Nishizaki et al. conducted a blood serum analysis for two Bull Sharks from the brackish (1.77 and 2.66 ppt) Usumacinta River basin. They found that, "the ionic concentrations of sodium, chloride and calcium were almost half that of the values for the seawater." Both specimens shared a higher ionic concentration of potassium (21% and 44% higher than that of seawater) but a drastically lower ionic concentration of magnesium (4% of seawater's). Urea content values were less than 30% of those recorded for marine Bull Sharks and 50%-70% of those recorded for Lake Nicaragua and Colorado River Bulls; however, no value was lower than that of the Potamotrygonids.

Potamotrygonid stingrays are exclusively freshwater elasmobranchs, and their adaptations can be seen as an ideal for freshwater success. Bull Sharks transitioning from saltwater to freshwater must actively decrease their ionic and urea concentrations in order to more closely match those of the adept Potamotrygonids. This process is very slow, and as such a Bull Shark's blood serum and urea content may, amazingly, be used as indicator for its freshwater residence time (lower concentration = longer residency).

This can explain the discrepancy in urea content values between the Usumacinta Bull Sharks and the other freshwater specimens: the Usumacinta sharks may have simply spent a longer time in their freshwater habitat than the Lake Nicaragua or Colorado Bulls.

1998

Freshwater Elasmobranchs from the Batang Hari Basin of Central Sumatra, Indonesia
Raffles Bulletin of Zoology, 46 (2): 425-429
TAN, H.H. & LIM, K.K.P.

Currently, there are three species of freshwater elasmobranchs known from the Batang Hari basin in Jambi, Sumatra. They are *Carcharhinus leucas* (Carcharhinidae), *Pristis microdon* (Pristidae) and *Himantura signifer* (Dasyatidae). *Pristis microdon* appears to be the rarest elasmobranch encountered from the basin. Both *Carcharhinus leucas* and *Himantura signifer* have economic value as food fish.

Carcharhnus leucas (Valenciennes)

Two small examples were bought from the fish market in Jambi during the dry season in July, 1997. The 790 mm (total length) male (Fig. 2) and the 810 mm female were reportedly caught from the Batang Hari.

The ability of bull sharks to ascend rivers and tolerate low salinity to fresh water for extended periods is well known. They have been recorded in the Aamzon basin some 4000 km inland, and have been known to breed in Lake Nicaragua (Last & Stevens, 1994: 245). There have been few records of bull sharks from inland localities in Southeast Asia. These include Malaysia (Sungai Perak, Sarawak), Vietnam (Dongnai), northern Java and the Philippines (Laguna de Bay, Argusan River, Saug River, Lake Naujan) (Compagno & Cook, 1995: 69). The present specimens apparently constitute the first record of this shark from within Sumatra.

The entirety of Tan et al.'s report on *Carcharhinus leucas* lies on the opposite page; there is little more that I can say about this shark in Sumatra. On the neighboring island of Java, however, the Bull Shark bears a legendary significance...

Long ago, only two powers lay claim to the rivers of Java: the crocodile, Baya, and the shark, Sura. They were great friends, but like all creatures, became greedy with hunger; at times of famine, the two powers fought mercilessly for control of the river and its resources.

On a hot day in the dry season, Baya spied a particularly choice goat on the riverbank and joyously declared it his. Upon hearing this, Sura became distressed, for he had not eaten in days; he then challenged Baya for the goat, and the two engaged in a tumultuous battle within the river. After hours of struggle, neither titan could triumph, and Sura, now exhausted, proposed the following to his wearied rival:

"Let us establish a truce and bounds: you are of the land, so you must eat only from the land, while I am of the sea, so I must eat only from the sea. We shall never compete for food again." Baya agreed to this reasonable treaty, and crawled up onto the riverfront, while Sura swam down into the ocean. The two never met for a very long time...

Until Sura became hungry once more: he could not find any food in the sea, and so swam upriver against his word. Baya, enraged at the sight of Sura breaking his truce, attacked. The crocodile and the shark exploded into their most violent battle yet, and each wrapped their jaws around the other's tail. Blood poured from both, but it was Sura who was the more injured; eventually, he could not endure any longer, and retreated from Baya's might. From that day forward, Sura never entered the rivers again.

This legend lives today in the statues and symbols of East Java's capital city: Surabaya.

117

1999

Mercury Levels in Four Species of Sharks from the Atlantic Coast of Florida
Fishery Bulletin, 97 (2): 372-379
ADAMS, D.H. & MCMICHAEL, R.H.

In May 1991, the Florida Department of Health and Rehabilitative Services (FHRS) released a health advisory urging limited consumption of all shark species from Florida waters. Owing to mercury concentrations in excess of U.S. Food and Drug Administration and State of Florida standards, FHRS recommended "adults should eat shark no more than once a week; children and women of childbearing age should eat shark no more than once a month." State of Florida guidelines recommended that fish containing less than 0.5 ppm of total mercury should represent no dietary risk, fish containing 0.5 to 1.5 ppm of total mercury should be consumed in limited amounts, and fish containing greater than 1.5 ppm of total mercury should not be consumed. The 1991 health advisory regarding sharks in Florida waters was derived from a limited number of samples taken from retail sources and from studies that lacked important information regarding species, capture location, sex, and size of sharks examined. Increased landings of sharks in Florida for human consumption (Brown, in press; FDEP) has prompted the need for more detailed information regarding mercury levels in Florida shark species.

Consequently, we report here analyses of total mercury levels in the muscle tissue of three carcharhinids (bull shark, *C. leucas*; blacktip shark, *C. limbatus*; and Atlantic sharpnose shark, *R. terraenovae*) and one sphyrnid (bonnethead, *S. tiburo*) from the east-central coast of Florida.

Bull Sharks can be mildly toxic to eat. This is, of course, not a unique attribute, as many species of shark, mackerel, tuna, and billfish are known to accumulate hazardous concentrations of mercury (primarily in the monomethyl form: CH_3Hg); but the very idea that the shark's flesh is poisonous—'cursed' to the human tongue—summons a heavier appreciation for the void between our species. Heterotrophy is, strangely enough, a powerful bridge between two organisms: the materials that make a human can be consumed and *incorporated* into a shark, and likewise, the materials that make a shark can be consumed and *incorporated* into a human. The elemental makeup that once composed a man could easily be transferred to a shark, and vice versa.

But with the introduction of poisons such as mercury, there comes resistance—a crack in the bridge. Large, predatory fish such as *C. leucas* cannot be so easily linked with the would-be human predator; the bonds that naturally unite such powers becomes increasingly strained, and large sharks like the Bull become more removed from our atomic cohesion. Thus, on even the smallest scale, the Bull Shark exhibits defiance—an unwillingness to cooperate with the human design on the subtlest of levels.

Adams & McMichael examined 53 neonate and juvenile Bull Sharks from the Indian River Lagoon region and found, "a significant positive correlation between total mercury level and bull shark length". This observation is a perfect reflection of mercury bioaccumulation: mercury doesn't quickly leave a shark's body, but instead lingers, growing in concentration every time a new mercury-bearing prey item is consumed. The Bull Sharks in this study exhibited an incredible range of mercury levels, from a benign minimum of 0.24 ppm to a toxic maximum of 1.7 ppm (mean concentration was 0.77 ppm). Such quantities are 'defiant' enough to recommend *against* shark consumption.

1999

Sélaciens du Miocène Terminal du Bassin d'Alvalade (Portugal) Essai de Synthèse

Ciências da Terra (UNL), 13: 115-129, 5 fig., 2 pl.

ANTUNES, M.T. & BALBINO, A.C. & CAPPETTA, H.

RÉSUMÉ

Mots-clés : Sélaciens; Miocène supérieur; Europe; Portugal; Paléoécologie

On présente la liste des taxa de Sélaciens reconnus d'après plus de 10.000 dents récoltées dans le Miocène terminal du basin d'Alvalade, Portugal. Il s'agit d'une faune très riche, ayant vécu dans la zone néritique, en climat tempéré chaud à subtropical et la plus modern d'âge miocène connue en Europe.

INTRODUCTION

La liste suivante dresse la repartition des taxa et le nombre de dents de Sélaciens identifies dans le basin d'Alvalade: gisements de Santa Margarida (S.M.), Esbarrondadoiro (Esb.) et Vale de Zebro (V.Z.) (BALBINO, 1995).

CARCHARHINIFORMES

Carcharhinidae

Carcharhinus cf. *leucas*	Esb. (7) V.Z. (4)
Carcharhinus cf. *obscurus*	Esb. (3)
Carcharhinus cf. *plumbeus*	Esb. (7) V.Z. (9)
Carcharhinus cf. *perezi* S.M. (25)	Esb. (380) V.Z. (160)

This is a study concerning the Miocene Elasmobranch fauna from the Alvalade Basin of Portugal, which is represented by fossil deposits in Santa Margarida (S.M.) Esbarrondadoiro (Esb.), and Vale de Zebro (V.Z.). Ancient sharks that may possibly link to the modern *Carcharhinus leucas* are quite rare in this assemblage; of about 10,000 tooth samples, *Carcharhinus* cf. *leucas* (with the "cf." indicating a species comparable to, though not necessarily the same as, the modern Bull Shark) is represented by only 11 teeth (or 0.1% of the entire sample). Given its rarity, little is said about *Carcharhinus* cf. *leucas* specifically, other than it was a coastal shark of tropical-subtropical climes.

Climatological distribution is a particular focus of Antunes et al., who attribute each represented species and genus to a broad climatological group: tropical/subtropical, temperate, and coldwater. The vast majority of the Alvalade Basin's 45 fossilized taxa are of the tropical/subtropical and temperate regimes; only four species (*Squalus* sp., *Pristiophorus* sp., *Raja olisiponensis*, and *Raja* sp.) belong to the coldwater group. Interestingly, these latter four species were nearly exclusive to the Esbarrondadoiro deposit, leading the authors to believe that Esbarrondadoiro may have been submerged in far deeper waters than Santa Margarida and Vale de Zebro during the Miocene (as greater depth would have accommodated for a greater stratification in temperature, wherein the warmer surface waters would overlie the deeper coldwater habitat; the Santa Margarida and Vale de Zebro sites would have been too shallow for stratification to develop).

Taxonomically, the Carcharhiniformes represent a majority in terms of the Alvalade Basin's species biodiversity (forming 40% of the total number of fossilized taxa). The Carcharhinidae are a particularly abundant family, and their success is a defining characteristic of this unique Miocene formation.

1999

An Index of Abundance for Juvenile Coastal Species of
Sharks from the Northeast Gulf of Mexico
Marine Fisheries Review, 61 (3): 37-45
CARLSON, J.K. & BRUSHER, J.H.

A fishery-independent assessment of juvenile coastal shark populations
in U.S. waters of the northeast Gulf of Mexico was conducted using two
methods: gillnets and longlines. Surveys were conducted monthly during
April-October in two fixed sampling areas from 1996 to 1998. The Atlantic
sharpnose shark, *Rhizoprionodon terraenovae*, and the blacktip shark,
Carcharhinus limbatus, were the most common species captured with either
longlines or gillnets. An additional 14 shark species were captured, and
juvenile indices of abundance were developed for 8 species with gillnets
and 6 species of sharks with longlines. Trends in catch-per-unit-effort were
found to vary depending on species. Length-frequency information revealed
that the majority of sharks captured were juveniles. Given the direct
relationship between stock and recruitment for sharks, continued
monitoring of juvenile abundance will aid in determining the strength of the
parental stock size and for predicting future population strength.

It is still unclear whether the abundance estimates presented herein
represent stockwide estimates or represent populations only for the
northeastern Gulf of Mexico. Although adults of many species, particularly
sandbar, blacktip, scalloped hammerhead, and spinner shark are highly
migratory, whether sharks from the eastern Gulf of Mexico mix with stocks
from the western Gulf of Mexico, Atlantic Ocean, or Mexican waters is yet
to be determined.

It appears as though the Bull Shark has earned a permanent place on the sidelines within American shark fisheries studies: in this particular Marine Fisheries Review publication, *Carcharhinus leucas* is (once again) hardly acknowledged, and only described as a species, "caught, but not consistently captured". To continue our discussion as to why the American Bull is so comparatively rare, let us examine its far more abundant cousins: the Sandbar (*Carcharhinus plumbeus*), the Blacktip (*Carcharhinus limbatus*), and the Atlantic Sharpnose (*Rhizoprionodon terraenovae*).

C. leucas chiefly differs from these three populous Carcharhinids in size, prey selection, and habitat preference. The Sandbar, Blacktip, and Sharpnose sharks are all comparatively smaller, and primarily consume fish and invertebrates; the larger *C. leucas* has a distinctly broader diet that includes reptiles and mammals (both marine and terrestrial). Though Sandbar, Blacktip, and Sharpnose sharks do have some tolerance for lower salinities, the Bull Shark is exclusive in its ability to exploit freshwater systems. All four sharks, however, share nearshore coastal habitat.

The longlines and gillnets used for American fisheries studies select very well for midwater (that is, not benthic) piscivorous sharks that inhabit the coastal euphotic zone; as such, their success with nearshore fish-eaters such as the Sandbar, Blacktip, and Sharpnose sharks goes without speculation. The lack of success with *Carcharhinus leucas* may simply be due to the species' larger variance in diet and habitat preference: rather than nabbing the small fish bait on longline hooks or swimming into a coastally deployed gillnet, the Bull may simply be spending more time chasing larger prey upriver.

Of course, this is all speculation, and there are many more variables to consider. Hopefully, we will discover the reason behind this relative rarity in data forthcoming.

1999

Genetic Identification of Sharks in the U.S. Atlantic Large Coastal Shark Fishery
Fishery Bulletin, 97 (1): 53-61
HEIST, E.J. & GOLD, J.R.

Nucleotide sequences of a 394-396 base pair fragment of mitochondrial (mt) DNA, including parts of the cytochrome *b* and threonine tRNA genes, were obtained for eleven species of carcharhiniform sharks important to the U.S. Atlantic large coastal shark fishery. Sequences were used to predict sizes of restriction fragments produced by 118 restriction enzymes with unique recognition sequences. Seven restriction enzymes were chosen that produce an array of species-specific fragments for the eleven species. Geographic variation was examined in several species by surveying specimens from geographically distant regions. Only one of the species, the spinner shark (*Carcharhinus brevipinna*), exhibited geographic variation in mtDNA restriction fragments. The sandbark shark (*C. plumbeus*) exhibited sequence polymorphism that did not produce differences in restriction patterns of any of the seven enzymes. We detected numerous differences between observed restriction patterns in ten tiger sharks (*Galeocerdo cuvier*) and patterns predicted from a published sequence. We concluded that the published sequence is incorrect. Amplification of a single PCR product from a sample of meat, digestion of aliquots of the product with restriction enzymes, and sizing of fragments on agarose gels is an efficient method for distinguishing among these eleven carcharhiniform sharks. The method can be applied when only a small amount of tissue is available.

At last, we have come to the subtlest and, arguably, most precise of the biological sciences: genetics. The power of this discipline is staggering: with only a small amount of tissue, a shark can be identified to the species (and in some cases, population) level. To understand how this is accomplished (especially in tandem with the material opposite), one must understand the labyrinth that is genetic knowledge. For our purposes, we need only to examine the broader impact of Heist & Gold's research.

This study's primary objective is to counter shark poaching (whether intentional or accidental). To effectively comply with shark fishing regulations and quota limits, fishermen must be able to positively identify the species captured; however, this is very difficult to do at sea, especially in the case of typical shark processing (where necessary diagnostic features, such as fin position, are removed via the finning, beheading, or gutting of the carcass). In response to this critical problem, Heist & Gold considered the utility of genetic discrepancies as diagnostic characters: specifically, they investigated the use of restriction-fragment differences in polymerase chain reaction amplified DNA as a, "rapid and inexpensive means of identifying carcasses and fins."

A restriction fragment is, in simpler terms, a unique piece of DNA 'cut' by a specific restriction enzyme. For a certain strand of mitochondrial DNA shared by 11 shark species, Heist & Gold employed seven restriction enzymes (*Alu*I, *Dde*I, *Fok*I, *Hae*III, *Hinc*II, *Hinf*I, *Rsa*I) to 'cut out' a characteristic arrangement of restriction fragments. This arrangement— called a 'restriction pattern'—varies between the 11 shark species, and can be an essential tool for identification. For Bull Sharks specifically, Heist & Gold analyzed seven individuals from the Florida Gulf Coast; all shared an identical restriction pattern on the species level, and exhibited no significant variance on the population level.

1999

Report to Parks Australia on Estuarine Fish Monitoring of
Kakadu National Park, Northern Australia
Australia MAGNT, Darwin: 51 pp.
LARSON, H.K.

The estuarine fishes of Kakadu National Park, which includes three
major rivers entering the Van Diemen Gulf in the Northern Territory, have
not been well documented (unlike the freshwater fishes of the region).
Midgley (1973) reported two estuarine fishes (a sole and a gudgeon of
unspecified species) from Cannon Hill Lagoon, which is "in effect a
tributary of the East Alligator River, and is subject to some tidal influence
during very high tides". Pollard (1974) detailed some fishes other than the
strictly freshwater species known from the East Alligator River system.
Bishop *et al.* (1981, 1990) added a few additional estuarine fishes. In total,
only 72 species of marine and estuarine fishes were definitely known from
the literature as occurring within Kakadu. Davis and May (1989) carried out
a nine-month survey of the East Alligator River, as part of a larger study on
barramundi.

The present knowledge of the fish fauna in the Park and surrounding
waters is based on these records, on prawn-trawl bycatch, demersal fish
trawl catch and a small amount of *ad hoc* collecting. Most of these
collections have been made outside the Park boundaries.

In 1996, the Ichthyology Section of MAGNT undertook a survey, as a
consultancy to Parks Australia North, of the residential estuarine fish fauna
of Kakadu. The survey took place from 24 May to 8 June 1997, with 93
collection stations made over 13 days.

According to this survey, the Carcharhinidae were found to be one of Kakadu National Park's most speciose estuarine fish families (placing 7th behind gobies, anchovies, fork-tailed catfish, mullet, jewfishes, and tongue soles). However, the evidence behind this claim contains a glaring mistake: *Nebrius ferrugineum* (today known as *Nebrius ferrugineus*) is a nurse shark (order Orectolobiformes, family Ginglymostomatidae) miscounted as one of this survey's seven Carcharhinids. This is most likely a small typing error, probably nothing more than a simple misalignment of data; however, the implications of such a flaw are profound. It's that pervasive question that haunts all platforms, be they scientific, religious, or political:

If I know that this statement is wrong, what other statements might be wrong?

No matter how small the slip, a misrepresentation of fact is a deeply corrosive chink in the armor of trust. We, as consciously existing beings — entities completely dependent on interpretation, assessment, and *understanding* of our universe for physical, mental, and spiritual survival — cannot bear to be misled; whether intentional or accidental, misinformation is a cancer to our faith in knowledgeable institutions. Science, government, and religion can guide us all into to a deeper, richer, and completely transformative understanding of our selves and our world; but when they fail us with misinformation, they corrupt us. They contribute to the Baseless Being—the human who may exert their power and resources in the misguided defense of something indefensible.

This specific example with *Nebrius ferrugineus* is small and largely inoffensive; but it is a mistake—a scientifically published mistake. When our empirical institutions fail to catch these mistakes and publish them as fact, they summon a very dangerous question: can we trust science? In an ideal world, we should never have to ask.

1999

Biodiversidade de Elasmobranquios do Brasil
Ministério do Meio Ambiente (MMA), Projeto de
Conservacao e Utilizacao Sustentavel da Diversidade
Biologica Brasileira (PROBIO), Recife 1999
LESSA, R.P. & SANTANA, F.M. & RINCÓN, G. & GADIG,
O.B.F. & EL-DEIR, A.C.A.

1. INTRODUÇÃO

A Classe Elasmobranchii é composta por peixes com o esqueleto
cartilaginoso: tubarões e raias. Os tubarões são distribuídos em todos os
mares e oceanos, em águas tropicais, subtropicais, temperadas e frias
apresentando hábitos demersais ou pelágicos (Compagno, 1984). Perfazem
um total de 8 ordens, divididas em 30 famílias com aproximadamente 370
espécies catalogadas. Por sua vez, as 500 espécies de raias atuais são
marinhas ou dulce-aqüicolas adaptadas à vida demersal ou pelágica (Last e
Stevens, 1994). Este levantamento identificou na costa brasileira 82
espécies descritas de tubarões e outras 3 espécies ainda por serem descritas
ou taxonomicamente revisadas. Dentre as raias, foram identificadas 45
espécies descritas e outras 6 também em processo de descrição ou revisão
taxonômica (*q.v. Anexo 1*).

Em seu ambiente natural, a maior ameaça a estes animais é a atividade
antrópica. Muitas populações de elasmobrânquios em todo o mundo estão
em depleção devido a quarto fatores: 1) a degradação dos ambientes
costeiros em que se desenvolvem; 2) a captura accidental (by-catch) e, nos
últimos anos, dirigida; 3) o aumento do esforço de pesca, e ainda pela 4)
estratégia de vida das espécies (Camhi *et al.*, 1998).

The Brazilian seacoast can be subdivided into five marine ecoregions:

- Amazonia (French Guiana border to Parnaíba)

- Northeastern Brazil (Parnaíba to Salvador)

- Eastern Brazil (Salvador to Cabo Frio)

- Southeastern Brazil (Cabo Frio to Criciúma)

- Rio Grande (Criciúma to Uruguay border).

According to this report composed by Lessa et al., *Carcharhinus leucas* appears to have a very significant presence in Brazilian waters (comprising more than 20% of the total coastal shark catch) and furthermore occurs in all five ecoregions (including the Amazon River). It is a target of both artisanal and industrial fisheries, and has been caught in Brazil with driftnets, gillnets, and longlines at depths between 2 and 65m.

Though shark fisheries management is by now a familiar subject, Lessa et al. tackle its provisions with delicate precision; I have never more clearly understood the 'perfect storm' that is the shark fisheries problem. In order for governments to commit to any shark species' management and protection, there must be extensive scientific research to justify such a commitment. However, this kind of research doesn't really exist for many shark species, because of their lower economic value: as sharks are not as commercially lucrative as teleosts like swordfish or tuna, they are often given a lower research priority.

Ignorant of priorities, sharks are biologically slow in growth, late in sexual maturity, and low in fecundity—the perfect ingredients for overexploitation. While other fish may receive strict management protection in the form of seasons, quotas, or moratoriums, many sharks are still 'left out in the open' for targeted fisheries, finning operations, or commercial bycatch. Simply put: the research is not matching the rate of depletion.

1999

*Tag Position and First Dorsal Fin Growth Relationships in
a Sphyrnnid and Nine Carcharhinid Sharks*
Journal of the Elisha Mitchell Sci. Society, 115 (1): 55-60
SCHWARTZ, F.J. & COOK, M. & MCCANN, C.

Dorsal fin one (D_1) and fork length (FL) body growth relationships were examined for young to adults of 129 male and 128 females representing nine carcharhinid and one sphyrnid sharks to note how they may affect tag retention. Females of seven species had higher $Dorsal_1$ fins than did males, except for blacktip males whose D_1 fins were always higher. Males of five species possessed longer dorsal fin bases as the shark grew. A tag placed at the base of the D_1 fin would remain in place. Tags placed in the middle of the D_1 fin would move upward in six species, laterally rearward in three species, and remain stationary in two species.

Rearward or upward movement of a tag inserted into the middle of the first dorsal fin will increase the likelihood of its being affected by the stream of water flowing along the surface of the fin (see Carrier, 1985; Olsen, 1953, Heupel et al., 1998, for examples). Although Heupel et al. (Fig. 4, 1998) felt roto tags were an efficient way to tag sharks, tearing away and skin damage was evident in response to the tag. Flowing water pressures, during fin and shark growth, will aid in its being pulled out, especially in silky, tiger, female blacknose, and other sharks.

Contrary to Davies and Joubert (1967), dart tags have little or no affect on shark growth (Schwartz, 1997). The base of the shark's first dorsal fin is recommended as the best tag retention site.

Sexual dimorphism is something we have yet to explore in regards to *Carcharhinus leucas*. We already understand the obvious differences — sperm versus eggs, testes versus ovaries, claspers versus no claspers — but what about the secondary sexual characteristics? Schwartz et al. tease one such detail: a fascinating discrepancy in first dorsal fin height between the sexes. For Atlantic Sharpnose (*Rhizoprionodon terraenovae*), Blacknose (*Carcharhinus acronotus*), Sandbar (*Carcharhinus plumbeus*), Dusky (*Carcharhinus obscurus*), and Tiger (*Galeocerdo cuvier*) sharks, females were consistently found to have taller first dorsal fins than males. In the case of Blacktip (*Carcharhinus limbatus*) sharks, the opposite was found to be true, but what about Bulls? Unfortunately, Schwartz et al. only sampled male *Carcharhinus leucas* for this study, so it is, as of yet, an unconfirmed possibility that there may be such a sexual difference.

Speculation aside, there is a confirmed equality in the rates of upward and lateral fin growth for *Carcharhinus leucas*. Most of Schwartz et al.'s sharks demonstrated a pattern of directional first dorsal fin growth, favoring either upward movement (Sandbar, Dusky) or lateral movement (Silky, Tiger). The male Bulls, however, shared the uncommon category of "equal directional fin growth" with Atlantic Sharpnose and male Spinner sharks. This pattern is important to understand in regards to shark tagging, especially in consideration of tag types placed in the center of the first dorsal fin (such as roto tags). Unlike dart tags (which are plunged securely into the fin base), these 'grow with' the shark; over time, they can be pushed outward from their initial tag site into more compromising locations on the first dorsal. Such movement puts the tag at greater risk of elemental exposure, increases the chance of dislodging, and furthermore acts to the detriment of the animal's health (given the confines of such a tag design).

2000

Marine Fishes from Ceará State, Brazil: I –
Elasmobranchii (Ictiofauna Marinha do Estado do Ceará,
Brasil: I. Elasmobranchii)
Arquivos de Ciências do Mar, 33: 127-132
GADIG, O.B.F. & BEZERRA, M.A. & FEITOSA, R.D. &
FURTADO-NETO, M.A.A.

Elasmobranch fishes from Ceará State's coast, Northeast Brazil, were studied under several aspects. According to the data obtained on this study, there are eight orders, 21 families, 27 genera and 42 elasmobranch species along the Ceará coast. Sharks comprise four orders (50% of the total elasmobranch from Ceará), 12 families (57.1%), 17 genera (62.9%) and 30 species (71.4%). Rays are represented by four orders (50%), nine families (42.9%), 10 genera (37.1%) and 12 species (28.6%). Coastal elasmobranch represent 38.1% of the total fauna. Oceanic species are represented by 28.6%, the oceanic/coastal are 33.3%, demersal elasmobranch represent 35.7%, pelagic are 71.4%, and reef species correspond to 42.2% of the total elasmobranch fauna of Ceará State. According to occurrence, four species are considered rare, 47.6% are frequent but not abundant, 42.8% are frequent and abundant, and 35.7% are migratory species. The population status is unknown for most elasmobranchs, corresponding to 71.4% of species. Such data suggest that future works should be dedicated to obtain more data on biological parameters on the main elasmobranch species targeted by the fisheries.

Key words: Elasmobranchs, sharks, rays, occurrence, Ceará State (Brazil).

This is simply a record for the Brazilian state of Ceará, which claims *Carcharhinus leucas* as an abundant resident plying the estuarine, reef, coastal, and pelagic habitats; the local population status of this species is unknown. Interestingly, Gadig et al. acknowledge a possible sampling bias in regards to the use of gill nets. Given that they are suspended in midwater, gillnets select against the capture of demersal species such as rays; the authors thus acknowledge that their classification of each species' abundance (FAB for abundant, FPA for uncommon, and ARR for rare) is possibly affected by gear type. Many (if not all) of the world's shark surveys are vulnerable to this kind of problem, as longlines, gillnets, and even traditional rods and reels select for fairly-sized epipelagic predators like the Bull, Mako, and Spinner sharks. Large planktivorous species such as the Whale, Basking, and Megamouth are impossible to tally in this fashion, while deepwater (Sixgill, Sleeper), demersal (Angel, Nurse), or small (Cat, Lantern) shark species remain largely out of the sampling reach (though they can be rarely caught).

I'm fascinated by the fact that *C. leucas* is once again named as an occupant of midwater habitat. Perhaps it's ignorance on my part, but I've grown accustomed to imaging this shark haunting the coastal shallows or the muddy river; indeed, the Bull's tiny eyes and freshwater plies betray a kinship with the surface sun. However, as documented both scientifically and on film, *C. leucas* can be found suspended within the water column, neither close to the sunny ceiling nor the sandy floor. It can be a shadow in blue—a dark materialization from a wholly enveloping tranquility. Its very presence disturbs the peace, and yet, it is unmistakably one with this environment, as fluid as the medium through which it transgresses.

We've entered into a bit of Romantic poetry— but how can we not, with such marvelous nature?

2000

Pelagic Sharks in the Indian Seas their Exploitation,
Trade, Management and Conservation
CMFRI Special Publication, 70: 1-95
PILLAI, P.P. & PARAKKAL, B.

Sharks play a critical role in the marine ecosystem as highly efficient predators which keep ecosystem population in check. However, the low reproductive potential of shark species make them vulnerable to overfishing thus making it imperative to take a more conservative approach concerning their commercial fishery. Further, the Convention on International Trade in Endangered Species (CITES) through a resolution in 1994 requested the FAO and other organizations to collect and collate biological and trade data on shark species. During the 1980s the shark fisheries were growing at a rapid pace fuelled by the demand for shark fin and shark meat, and currently shark fisheries cover the entire world oceans.

Though a harvestable potential of 168,000 t of elasmobranchs has been estimated from the EEZ of India (MOA, GOI, 1991), they are not fully exploited as evident from the average production of elasmobranchs (61,591 t) and that of sharks (41,483 t). Pelagic sharks constitute about 68% of the total shark landings in India indicating therein scope for expanding the commercial exploitation of this group of fishes which has to be implemented in a planned manner without affecting their population.

This is one of the clearest and most detailed accounts of the global shark fishery's progression from amorphous origins to alarming excess. In the 1950's and early 1960's, sharks were largely disregarded due to their repulsive taste and odor (resulting from high concentrations of urea and trymethylamine in their flesh). Large sharks, like *Carcharhinus*, were primarily landed by longline and drift gillnets prior to the 1960's, while smaller species were caught in seines. However, the method of harvest largely changed with the introduction of motorization in the 60's and 70's.

While motorized trawls and longlines increased shark fishing efficiency, a separate development increased incentive: the rise of market fish prices. As traditional food fishes were becoming more expensive during the 1980's, consumer attention shifted towards sharks as source of inexpensive protein. This shift in demand came in tandem with a reduction of tariff rates for imported shark fins (which, by this time, fetched a considerably high price). Simultaneously, the tuna longline fishery—an important source of shark bycatch—experienced dramatic growth throughout the 1980's as well.

Given such compelling variables, the 80's shark fishery exploded. Flesh, fins, liver, oil, skin, and cartilage saw a boost in demand, and held a valuable position throughout the 1990's. It is at this time that concerns began to rise, and India, like most shark-harvesting countries, noticed an alarming trend: the steady decrease in the average length of landed sharks — a, "clear indication that over exploitation is beginning to leave a telling effect". The Convention on International Trade in Endangered Species (CITES) established a resolution in 1994 to, "collect and collate necessary information on biological and trade data on sharks." This effort resonates today within the global modern endeavor to synergize for the critical benefit of effective shark fisheries management.

2000

Elasmobranchs of the Cape Fear River, North Carolina
Journal of the Elisha Mitchell Sci. Society, 116 (3): 206-224
SCHWARTZ, F.J.

Twenty-four species of elasmobranchs (sharks, skates, and rays) were collected during an intensive six-year (1973-1978) survey of 22 stations located throughout the Cape Fear River estuary, Carolina Beach Inlet, and the adjacent Atlantic Ocean. Three additional species (whale shark, *Rhincodon typus*, basking shark, *Cetorhinus maximus*, and spinner shark, *Carcharhinus brevipinna*,) reported in the literature, brought the elasmobranch fauna to 27 species. Elasmobranchs entered the Cape Fear River estuary from the Atlantic Ocean at its mouth near Southport, North Carolina, or via Carolina Beach Inlet. Channel, east and west shoal stations were sampled using otter trawls and gill nets from January through November of each year. Water temperatures, salinities, and dissolved oxygen levels were collected at each station in the river ecosystem. Three species dominated the elasmobranch fauna: Atlantic sharpnose shark (*Rhizoprionodon terraenovae*), Atlantic stingray (*Dasyatis sabina*), and clearnose skate (*Raja eglanteria*). Freshwater inflows or marine intrusions affected various elasmobranch use and penetration of the Cape Fear ecosystem. Most specimens were females. *R. eglanteria* and *D. sabina* were the first Elasmobranchs to enter the Cape Fear system each January or February; *D. Sabina* was the last species to leave in November. Many species occurred where river waters were warm, whereas only a few occurred when waters were cool.

136

The Cape Fear River is familiar to me; I was there, once, at the mouth. I arrived in early June 2015 and left the day of the second North Carolina shark attack on Oak Island. I could see the Oak Island lighthouse every night.

From what little I learned of Cape Fear, I saw sunsets, starlight, and warmth. I met two sharks in the estuary; a small Carcharhinid caught by a boy, and a Bonnethead who stalked me in the shallows. The latter was a very special encounter; there were no hooks, tanks, or diving masks. It was just two beings existing as nature intended.

The shallows were clear, and the baking seabed was an inviting mix of orange and tan. But I didn't go deeper. Appropriately, it was because of fear—but not necessarily of the emerging news. It was the fear of the unknown—the fear that stirs me in regards to all of the dangerous sharks. I knew that *Carcharhinus leucas* was in this estuary, but I didn't know when. I'll never know when. At the time of my play with the Bonnethead, there could have been a Bull three yards away; or, perhaps more appropriately, a thousand yards away. I'll never know.

Two of Schwartz's Bull Sharks were caught in September 1978 near Horseshoe Shoal, upriver from Southport and across from the North Carolina Aquarium at Fort Fisher. The water temperature was 26°C and the salinity was 23 ppt. Dissolved Oxygen—a measure we have beforehand never encountered—was 6.6 ppm (a fairly healthy, though not ideal, reading: under 5.0 ppm is stressful, while over 7.0 ppm is healthy enough for growth).

To get to Horseshoe Shoal, the Bulls most likely entered at the river mouth—at a point within eyesight of my Bonnethead rendezvous. How strange it is to imagine: that within the spheres of our existence, we have touched the same body of sea.

2001

Guía de Identificación de Peces Marinos de Nicaragua
**Proyecto para el Desarrollo Integral de la Pesca Artesanal
en la Región Autónoma Atlántico Sur, Nicaragua
COTTO, S.**

Las aguas del Caribe se caracterizan por poseer una elevada diversidad en lo referente a fauna marina, hasta la fecha se han encontrado 1,830 especies de peces, de los cuales 786 son considerados comerciales, o sea, aptos para consumo, de estos, en nuestro país se tienen identificados y clasificados 308 peces, de los cuales se han seleccionado los más abundantes tanto en las capturas como en los viajes de investigación para ofrecerlos bajo la forma de un manual de campo que facilite la identificación rápida y certera de aquellos de mayor interés que habitan en su gran mayoría en la zona de manglares, zona costera y en el área de la plataforma continental.

Para la preparación de este documento se tomó como base el "Listado Taxonómico de los peces identificados en los océanos Pacífico y Atlántico de Nicaragua" (Cotto A. 1998), sobre la base que contiene una relación de todas las especies exitosamente identificadas en ambas costas del país, así mismo, par alas descripciones se contó con el auxilio de variada literature especializada, con especial énfasis en las guías de Cervigón, la synopsis de Guitart y la más reciente publicación de Humman para los peces de arredife del Caribe.

Nicaragua is a place of brilliant biodiversity, but comparatively inchoate research; not many local fisheries scientists are available to collaborate and corroborate information regarding the dazzling array Nicaragua's Caribbean and Pacific species. Cotto's Caribbean field guide is in itself a call to further academic development: it currently describes 308 of the most abundant species in brevity, but also encourages readers to directly email the author with empirical updates. Like our previous FAO guides, this account is meant to spur dialogue amongst fishermen and scientists alike, for the sake of efficient communication and shared wildlife enlightenment.

The common trend to claim *Carcharhinus leucas* as the "most dangerous of all tropical sharks" is reinforced in this account; but is this claim accurate? Consider the International Shark Attack File and its spectrum of values (incidences of attack) regarding the four species of 'maneating' shark. As of January 19th, 2016, they were:

	Non-fatal Unprovoked	Fatal Unprovoked	Total
Bull Shark: *C. leucas*	73	27	100
Tiger Shark: *G. cuvier*	80	31	111
Great White: *C. carcharias*	234	80	314
Whitetip: *C. longimanus*	7	3	10

All four species enter the tropics, but only the Bull and Tiger could be fairly considered 'tropical' (i.e., 'restricted' to warmer waters). Even so, these two species have comparable values, making Tiger Sharks a worthy contender for the title of 'most dangerous' in the tropical realm.

2001

Field Guide to Requiem Sharks (Elasmobranchiomorphi:
Carcharhinidae) of the Western North Atlantic
NOAA Technical Report NMFS, 153: 1-36
GRACE, M.

Identification problems are common for many sharks due to a
general lack of meristic characteristics that are typically useful for
separating species. Other than number of vertebrae and number and shape
of teeth, identifications are frequently based on external features that are
often shared among species. Identification problems in the field are most
prevalent when live specimens are captured and releasing them with a
minimum of stress is a priority (e.g., shark tagging programs).
Identifications must be accurate and conducted quickly but this can be
challenging, especially if specimens are very active or too large to be
landed without physical damage. This field guide was designed primarily
for use during field studies and presents a simplified method for identifying
the 21 species of western North Atlantic Ocean sharks belonging to the
family Carcharhinidae (carcharhinids). To assist with identifications a
dichotomous key to Carcharhinidae was developed, and for the more
problematic *Carcharhinus* species (12 species), separation sheets based on
important distinguishing features were constructed. Descriptive text and
illustrations provided in the species accounts were developed from field
observations, photographs, and published references.

Bull shark, *Carcharhinus leucas* **(Fig. 14)**
Distinguished from sandbar sharks by lack of an interdorsal ridge.

Let's break down the Bull Shark's NOAA-defined field diagnostics by keying this species out from the Western North Atlantic Carcharhinidae (**21** species total). The Bull Shark's first dorsal fin is positioned more closely to the pectoral fin axil (that 'notch' formed between the body and the fin's inner margin) than the pelvic fin origin (**20** species left; Blue Shark eliminated). Like most Carcharhinids, the Bull Shark has no caudal keels (**19** species; Tiger Shark eliminated). Its second dorsal fin is smaller than its first (**18** species; Lemon Shark) and anterior to the, "anal fin base midpoint" (**14** species).

The Bull Shark's broad snout easily separates it from the Daggernose Shark (**13**). Sharply rounded pectoral and first dorsal fin tips just as easily distinguish this species from the Oceanic Whitetip (**12**). At this point in the NOAA key, we come to the most important character: the lack of an interdorsal ridge. Only five of the twelve remaining sharks have this trait: the Bull, Spinner (*C. brevipinna*), Blacknose (*C. acronotus*), Blacktip (*C. limbatus*) and Finetooth (*C. isodon*) sharks (**5**).

To readily separate *Carcharhinus leucas* from these four others, we must reexamine the first dorsal fin's positioning. For Bull Sharks, this fin's origin will always be over (or anterior to) the pectoral fin's inner margin, but never posterior. Spinner and Blacknose sharks can be usually keyed out at this point, but not always. Unpigmented (or dusky) fin tips can provide the final confirmation against the Spinner or the Blacktip (**3**), while long, broad pectorals can do the same against the Blacknose or the Finetooth (**1**).

Though we can be nearly certain of *Carcharhinus leucas* at this point, NOAA recommends an additional check for three essential characters: a short and bluntly rounded snout; a high and triangular first dorsal fin; and upper teeth that are broad and serrated. These serve as the final confirmation of a Western North Atlantic Bull Shark.

2001

The Neogene Sharks, Rays, and Bony Fishes from Lee Creek Mine, Aurora, North Carolina
In: Geology and Paleontology of the Lee Creek Mine, North Carolina, III, Clayton E. RAY & David J. BOHASKA.
Smithsonian Contributions to Paleobiology,
90: 71-202, 84fig., 1 tabl.
PURDY, R.W. & SCHNEIDER, V.P. & APPLEGATE, S.P. &
MCLELLAN, J.H. & MEYER, R.L. & SLAUGHTER, R.

The fish remains, including 104 species from 52 families, collected at the Lee Creek Mine near Aurora, Beaufort County, North Carolina, constitute the largest fossil marine fish assemblages known from the Coastal Plain of the eastern United States. The fish faunas came principally from the Pungo River Formation (Burdigalian, planktonic foraminifera zones N6-7) and the Yorktown Formation (Zanclian, planktonic foraminifera zone N18 and younger). A few specimens were obtained from the James City Formation (early-middle Pleistocene).

As an assemblage, the fishes found in the Pungo River Formation, including 44 species of selachians and 10 species of teleosts, are most similar to those from the "Muschelsandstein" of the Swiss Molasse.

The Yorktown Formation fish assemblage includes 37 species of selachians and 40 species of teleosts, derived mostly from the base of the Sunken Meadow Member.

Although the Pungo River Formation fish fauna is dominated by warm-water (18°-25°C) taxa, the Yorktown Formation fossil fish fauna includes warm and cool water species. Both fish assemblages occur with a cool-temperate invertebrate fauna.

The sands of time run white with bone, whose record marks once-life
unknown.

Fossils have such a power to explode our perspective; as suggested by this record, *Carcharhinus leucas* has essentially maintained its modern physique for at least 5 million years (and perhaps even longer). Fossilized *C. leucas* teeth found in the Lee Creek Mine's Yorktown Formation (Zanclian, Pliocene, 5.33-3.60 Ma) and Pungo River Formation (Burdigalian: Miocene, 20.44-15.97 Ma) are, according to the authors, "identical to those of the extant species." This is incredible: how can an animal be so stable for so long? Even in the case of only one feature (the teeth), it is astounding that a 5 (plus) million-year-old phenotype could be so unchanged—that scientists can precisely match it to that of a modern shark on the species level.

Like its contemporary descendant, the 'Ancestral Bull' possessed broad, equilateral, and finely serrated upper teeth, whose size rivaled both the ancient and modern *C. longimanus* (Oceanic Whitetip) as the largest of the entire genus. Such massive teeth were (and still are) appropriate for a highly varied and opportunistic diet of: bony fish (including tarpon, sea catfish, tuna, sea bass, and bluefish), sharks, rays, invertebrates, seabirds, sea turtles, and even, shockingly, whales.

The Neogene *Carcharhinus leucas* may be possibly linked to (or synonymous with) a variety of fossil species, including *Corax egertoni*, *Prionodon egertoni*, and *Pterolamiops longimanus*. *Carcharhinus leucas* teeth are also fascinatingly similar to those of *C. longimanus* in both size and shape; could that hint to a closer phylogenetic relationship? As affirmed by the authors, many more samples need to be collected before this—or even a general phylogeny of the fossilized *Carcharhinus*—could be investigated.

2002

Shark Nurseries in the Northeastern Gulf of Mexico
**In: McCandless CT, Pratt HL Jr, Kohler NE (eds) Shark
Nursery Grounds of the Gulf of Mexico and East Coast
Waters of the United States: An Overview. An Internal
Report to NOAA's Highly Migratory Species Office. NOAA
Fisheries, Narragansett, RI: 165-182**
CARLSON, J.K.

Scope

Sharks were sampled by the National Marine Fisheries Service, Panama City Laboratory from March 1993-October 2000 as part of various studies on shark population dynamics and life history. All studies were directed towards sharks but focused on establishing a fishery independent index of abundance in the northeastern Gulf of Mexico; collecting information on age, growth, and reproduction; longline and gillnet selectivity; and feeding ecology. All studies were funded by NMFS/Highly Migratory Species Office, Washington D.C.; Southeast Fisheries Science Center's Sustainable Fisheries Division; and the NMFS Panama City Facility.

Bull Shark, *Carcharhinus leucas*

Of all species collected, bull sharks seemed to prefer the most particular habitat type (Table 12). Bull sharks (n=36) ranging in size from 65-267 cm TL were only collected in areas with silt/clay sediment, high volumes of freshwater inflow, and high turbidities (water clarity from 66-103 cm). This species was found in water temperatures between 20.7-31.8°C, salinities of 25-36 ppt and depths of 2.5-5.3 m.

This study considers five major bay systems in the Northeastern Gulf of Mexico: Apalachee Bay, Apalachicola Bay, St. Joseph Bay, Crooked Island Sound, and St. Andrew Bay. Of these, Apalachicola Bay yielded the highest number of captured Bull Sharks (31); the other four systems only produced one or two individuals each.

Considering this study's scope, the total number of captured Bull Sharks (36) is neither impressive nor warranting of significance; this species is, once again, proportionally rare in the classic American coastal fisheries study. However, the discrepancy of captures between bay systems is rather curious; why did the Apalachicola system yield a higher amount of Bull Sharks than the other bays?

Carlson speculates that Apalachicola Bay may simply be more accessible to large predators (like the Bull Shark) than the other systems. This is inferred from an interesting observation: that shark species with larger juveniles and young-of-the-year (such as the Blacktip, Spinner, Sandbar, and Finetooth) were found more often in Apalachicola, while species with smaller juveniles and young-of-the-year (such as the Atlantic Sharpnose, Bonnethead, and Blacknose) were found more often in Crooked Island Sound, as well as the shallower areas of St. Joseph Bay and Apalachee Bay. Unlike Apalachicola, these latter systems are smaller, shallower, and more enclosed.

From this interesting distributional pattern, Carlson speculates that the latter systems might be used as shark nurseries. Given their physical structure and relative lack of large predators, these habitats may be simply more restrictive than Apalachicola as a foraging ground for *Carcharhinus leucas* (thus explaining the abundance a typical prey item: sharks with smaller neonates and young of the year). More demographic research is, of course, needed before we can fully commit to this promising idea.

2002

Field Identification Guide to Western Australian Sharks and Shark-Like Rays
Fisheries Occasional Publications, 1: 1-25
MCAULEY, R.B. & NEWBOUND, D.R. & ASHWORTH, R.

Sharks and their relatives (the skates, rays and chimeras) are a highly diverse group of fish that evolved over 400 million years ago. These fish (collectively called Chondrichthyes) are characterized by a cartilaginous skeleton; multiple gill openings; skin covered with modified teeth instead of scales and external male reproductive organs. Over 160 species of sharks are known to inhabit Australian seas, although new species continue to be discovered. Sharks have been so evolutionarily successful that they inhabit all aquatic habitats: from freshwater rivers and lakes to ocean depths of thousands of metres.

As 'apex predators', many shark species occupy the very top level of the food chain and thereby play an essential role in maintaining the health of the marine environment. As well as their environmental importance, sharks provide a valuable resource for both the fishing and tourism industries, and chemical compounds derived from shark products are being examined for their potential pharmaceutical uses, particularly for cancer and arthritis treatments.

Despite their significance, sharks are a poorly understood group which urgently require further scientific study. As a first step, this guide is intended to improve the standard of identification and shark-catch reporting in Western Australia's widespread and varied fisheries.

Sharks get cancer. Shark cartilage cannot cure cancer. These truths are slowly becoming more apparent to the modern public, but in 2002, that was not the case. Mcauley et al.'s trivial remark that, "chemical compounds derived from shark products are being examined for their potential pharmaceutical uses, particularly for cancer and arthritis treatments" is a fascinating snapshot of the time's scientific and cultural mindset. The myth that sharks were immune to virtually all diseases was still pervasive within the public (and scientific) psyches, but this mindest has been almost exclusively based on pseudoscience (at best). How did this happen?

In the 1970s, medical researchers discovered anti-angiogenic properties in mammalian cartilage—properties that could prevent tumors from developing. They speculated (though never confirmed) that such properties would most likely be present, nay, abundant, in Selachians, as their entire skeletons are cartilaginous. Given that most people had never seen or heard of a shark with cancer at that time, the idea was appealing. Unfortunately, nobody seemed aware of the fact that shark tumors have already been recorded since 1908 (to be fair, such records were scarce).

By 1992, the myth's position in popular culture became solidified with Dr. I. William Lane's now-infamous bestseller, *Sharks Don't Get Cancer: How Shark Cartilage Could Save Your Life*. Riding off this successful deception, Dr. Lane established the pharmaceutical company LaneLabs, which began to market its own 'cancer-fighting' cartilage pill, BeneFin. This and other hack cartilage products have directly increased the now-disastrous global shark fishing effort. Fortunately, the myth's foundations began to crack in the late 1990s and early 2000s with lawsuits and more thorough research. By 2004, Dr. Gary Ostrander provided a deafening blow with his published (and publicized) description of shark tumors, which included malignant forms.

2003

The Conservation Status of Australasian Chondrichthyans
**Report of the IUCN Shark Specialist Group Australia and
Oceania Regional Red List Workshop**
The UQ, School of Biomedical Sciences, Brisbane, Australia
CAVANAGH, R.D. & KYNE, P.M. & FOWLER, S.L. &
MUSICK, J.A. & BENNETT, M.B.

IUCN – the World Conservation Union, is the umbrella body for
the world's conservation agencies and institutions. Its members comprise
sovereign states, government agencies and non-governmental organisations.
The Species Survival Commission (SSC) is a volunteer network within
IUCN, comprised of nearly 7,000 scientists, field researchers, government
officials and conservation leaders from almost every country in the world,
and is an unmatched source of information on biological diversity and its
conservation. As such, SSC members provide technical and scientific
counsel for conservation projects throughout the world and serve as
resources to governments, international conventions and conservation
organisations.

In response to growing awareness and concern of the severe impact
of fisheries on elasmobranch populations around the world, the SSC
established the IUCN Shark Specialist Group (SSG) in 1991; it is now one
of the largest and most active specialist groups within the SSC. The SSG
provides leadership for the conservation of threatened species and
populations of all chondrichthyan fishes. It aims to promote the long-term
conservation of the world's sharks and related species (the skates, rays and
chimaeras), effective management of their fisheries and habitats and, where
necessary, the recovery of their populations.

148

The IUCN Red List is an immensely powerful shark conservation tool, as it establishes, "a baseline against which to monitor future changes in the global and regional status of chondrichthyan fishes." It is arguably the world's most comprehensive consolidation of chondrichthyan research, and its reports (updated yearly) are immediately accessible to the general public (via the IUCN Red List website, iucnredlist.org). This effective degree of international cooperation is essential to shark conservation, research, and management efforts; very few species can be satisfied by a single government's protection, as most are migratory, wide-ranging, or otherwise plagued by multilateral efforts beyond a single region's control.

The IUCN Red List applies one of nine assessment categories to each of its reviewed species: EX (Extinct), EW (Extinct in the Wild), CR (Critically Endangered), EN (Endangered), VU (Vulnerable), NT (Near Threatened), LC (Least Concern), DD (Data Deficient), and NE (Not Evaluated). The categories EX, EW, CR, EN, and VU are considered to be of the 'threatened' group; categories NT and LC are deemed comparatively 'safe', while DD and NE are applied to more cryptic species lacking the research needed for a proper conservation assessment.

Carcharhinus leucas is classified as globally Near Threatened (NT). The Bull Shark is common, and rarely targeted by fisheries, but its occurrence in estuaries and freshwater, "makes it very vulnerable to human impacts and habitat modification." This factor, in addition to a typically slow chondrichthyan life history, raises some concern for the species' wellbeing. Assessment authors Colin A. Simpfendorfer and George H. Burgess specifically mention that the, "average length of bull sharks caught by the Natal Sharks Board have declined significantly." This is a particularly disturbing omen.

2003

Application of DNA-based Techniques for the
Identification of Whaler Sharks (Carcharhinus spp.)
Caught in Protective Beach Meshing and by Recreational
Fisheries off the Coast of New South Wales
Fishery Bulletin, 101: 910-914
CHAN, R.W.K. & DIXON, P.I. & PEPPERELL, J.G. & REID,
D.D.

The International Union for the Conservation of Nature's (IUCN)
development of the Shark Specialist Group is indicative of the increasing
environmental awareness of sharks' crucial ecological role as apex
predators and that they are being threatened by human activities. Although
the conservation status of certain carcharhinid species (*Carcharhinus
limbatus*, *C. obscurus*, and *C. plumbeus*) are presently considered at low
risk or near threatened according to the IUCN's threatened species
categories,[1] species from the genus *Carcharhinus* are known to inhabit the
waters of New South Wales (NSW), Australia (Stevens, 1984; Last and
Stevens, 1994); however their conservation status has not been determined.
Known as whaler or "requiem" sharks, they are also commonly caught off
the coast of New South Wales in commercial fisheries (Stevens and
Wayte[2]; Tanner and Liggins[3]), recreational fisheries (Pepperell, 1992;
Gartside et al., 1999; Steffe et al.[4]) and by protective beach meshing (Reid
and Krogh, 1992; Dudley, 1997).

Because of morphological similarities between a number of shark
species in the genus *Carcharhinus* (Last and Stevens, 1994; Naylor and
Marcus, 1994), taxonomic identification to species level has been difficult
or inaccurate (or both)(Stevens and Wayte[2]).

150

Inaccurate identification is a serious problem for any conservation effort, leading to either improper species management, or to no management at all. Regional species assessments for the *Carcharhinus* of New South Wales have been difficult to produce, as sharks caught in beach meshing or recreational fishing have rarely been identified to the species level. This lack of species-specific data is challenging for conservationists to implement (and enforce) confident management policy; if a vulnerable species can't be identified at the dock, how can it be protected?

Genetic techniques yield an answer to this critical issue: provided with only a small tissue sample, methods such as PCR-based restriction fragment length polymorphism (RFLP), DNA sequencing, isoelectric focusing of muscle proteins, and direct multiplex PCR amplification can be powerful tools to unlocking a species' genetic 'fingerprint'. In this study, Chan et al. detail the 'fingerprints' of six *Carcharhinus*: *C. limbatus*, *C. brachyurus*, *C. leucas*, *C. obscurus*, *C. falciformis*, and *C. brevipinna*.

To create a genetic profile unique to each species, the authors first amplified a piece of mitochondrial DNA with a polymerase chain reaction (PCR): specifically, the, "1146 nucleotide base-pair (bp) cytochrome *b* (cyt *b*) region of the mtDNA". After PCR, Chan et al. subjected the cyt *b* fragments to restriction fragment length polymorphism (RFLP) with five restriction enzymes: *Alu* I, *Hae* III, *Pst* I, *Taq* I, and *Xho* I. Through this latter technique, each shark's cyt *b* fragment became 'cleaved' in a characteristic pattern. It is this pattern that provides the profile—the 'fingerprint'—as it can only be replicated within a single species; no two *Carcharhinus* will have the same cty *b* RFLP pattern.

It is wondrous to think of *Carcharhinus leucas* as not only a physical shark, but also a tangible code. These two unique identities—the genetic and phenotypic— are mirrors...in a sense, nature's ultimate taijitu.

2003

Stomach Contents of Pelagic Sharks in the Eastern Pacific Ocean Abstract
AES 19th Annual Meeting, Manaus, Brazil
GALVAN-MAGAÑA, F. & OLSON, R.J.

The bycatch of large predators, including sharks is common in the tuna purse-seiner fishing in the Eastern Pacific Ocean. We analyze 508 stomach contents from 6 shark species and three shark groups, which were sampled at sea by observers of the Inter-American Tropical Tuna Commission (IATTC) aboard of vessels from Colombia, Mexico, Panama, and Venezuela. The purse-seine sets yielding the samples were distributed across the geographical range of the EPO tuna fishery during December 1992 through September 1994. Our results including that Blue shark (*Prionace glauca*) predate mainly on cephalopods (decapods), *Argonauta* spp. and *Onychoteuthis banksii*. The Bull shark (*Carcharhinus leucas*) feed on *Auxis* spp. and cephalopods. Silky shark (*C. falciformis*) predates on *Engraulis mordax*, *Cubiceps pauciradiatus* and *Decapterus* spp. The mako shark (*Isurus oxyrinchus*) consumes *Abraliopsis* spp., *Acanthocybium solandrii*, *Auxis* spp. and *Euthynnus lineatus*. The white nose (*Nasolamia velox*) feed on *Katsuwonus pelamis* and *Thunnus albacares*; whereas the oceanic white tip (*C. longimanus*) predate on *C. pauciradiatus* and *Stenoteuthis oualaniensis*. The group of hammerhead sharks feed on cephalopods (*S. oualaniensis*, *Dosidicus gigas*, and *Abraliopsis falco*. The group of thresher sharks consumes on epipelagic fishes as *E. mordax*, *C. pauciradiatus* and mesopelagic fishes as *Benthosema panamense*. The carcharhinid group feed mainly on *E. mordax*.

This abstract is to one of many oral presentations showcased at the American Elasmobranch Society's annual meeting on elasmobranch science, conservation, and education; due to the verbal presentation format, its full transcription is not publically available. However, this abstract is in itself a crisp (albeit, partial) snapshot of the pelagic food web of the Eastern Pacific. We can see the bedrock of this web in *Engrarulis mordax*—a species of anchovy. This filter feeder is a crucial bridge between energy rich plankton and larger predators, such as tuna (*Thunnus albacares*, *Katsuwonus pelamis*, and *Euthynnus lineatus*), wahoo (*Acanthocybium solandrii*), and the impressive variety of documented squid species (*Onychoteuthis banksii*, *Stenoteuthis oualaniensis*, *Abraliopsis falco*, and *Dosidicus gigas*). Alongside scads (*Decapterus spp.*) and driftfish (*Cubiceps pauciradiatus*), these squid and mackerel form a rich forage base for the requiem, hammerhead, thresher, and mackerel sharks that Galvan-Magaña et al. detail.

The pelagic nature of this forage base (which also includes some deepwater denizens like the lanternfish, *Benthosema panamense*, and the Humboldt squid, *Dosidicus gigas*) fits perfectly in line with the purse seine's sampling design. Stretching to a maximum depth of 200m (the edge of the sunlit zone) and to a maximum length of 2,000m, purse seines are essentially 'walls' of netting designed to successfully encircle large schools of pelagic fish or squid. Once a purse seine surrounds a target school, its area gradually shrinks (as the netting becomes pulled-in by the lead line), and its bottom 'purses' (preventing the school from escaping downward). As one can imagine, this sampling design is potent, but monstrously ineffective in selecting against unwanted predators such as sharks, marine mammals, and sea turtles 'caught in the fray'. Pelagic Bull Sharks seeking frigate tuna (*Auxis* spp.) and cephalopods are among these casualties.

2003

Habitat Use by Three Shark Species in Charlotte Harbor, Florida Abstract
American Elasmobranch Society 19th Annual Meeting, Manaus, Brazil
HEUPEL, M.R.

A series of 19 acoustic hydrophones were placed in lower Pine Island Sound, Charlotte Harbor, Florida, to monitor the movement patterns of three shark species. A total of 58 blacktip (*Carcharhinus limbatus*), bonnethead (*Sphyrna tiburo*) and bull (*Carcharhinus leucas*) sharks were collected within the study site and fitted with acoustic transmitters for long-term monitoring efforts. Twenty-six young-of-the-year blacktip sharks were monitored for periods of 1-157 days. Twenty-one bonnethead sharks varying in size and age were monitored for periods of 1-104 days. Eleven bull sharks (125-183 cm TL) were monitored for periods of 3-52 days. Examinaton of movement patterns of the three species was used to define any overlap or differences in habitat use. Blacktip and bonnethead sharks appeared to use smaller activity spaces and be more resident within the study region than bull sharks. Bull sharks moved into and out of the study region regularly and overlapped their habitat use with those areas used by blacktip and bonnethead sharks. Movement and behavior patterns observed for all three species will be discussed.

Let us imagine Pine Island Sound, enclosed by land but embedded with sea. Looking down, the water is a bright olive, with shadows of seagrass drifting under sorbet twilight. The air is thick and warm— uncomfortably close, but improving with the emergence of a full moon. Mangroves ornately fringe the shorelines, acting as broaches to the water's breasts—the rising and falling breaths of tide, lapping our ears with caressing collapses. Saturation is the name of this scene—the alluring, pensive, and seductively fruitful energy of our mind's eye. Within this composition of riches, we embrace three permeances: the exuberant Bonnethead (*Sphyrna tiburo*), the copious Blacktip (*Carcharhinus limbatus*), and the corpulent Bull (*Carcharhinus leucas*).

The eventide begins with the Bonnetheads' attentive probing. Spasmodic twists and turns indicate an active hunt: though crab, fish, and squid are enveloped within the seagrass meadow, they are not protected, as their small but tactful predator courses through the rippling shallows. A young Blacktip, only slightly larger in size, simultaneously darts into the mangrove caves nearby, searching for jacks, grunts, and snapper. It takes no notice of its larger siblings, who meanwhile careen into whirling balls of menhaden and mullet dancing in the Sound's more denuded tracts.

By such electricity, the Bull Shark is piqued; from a larger void, it emerges, penetrating the liturgical ballad with a hulking, but silent, bearing. Its attitude is fickle, curious, but concealed; to the denizens of Pine Island Sound, it is but a resolute and infrequent shadow. Why does it enter the meadow tonight? Will it find success, and be provided? Or will it remain a shadow, passing over until the unexpected second coming?

Beyond science, we have dreams. Beyond empiricism, we have imagination. This is the essence of art: to reach for fantastic boughs, with feet submerged in reality.

2003

Observations on the Skeleton of the Heterocercal Tail of
Sharks (Chondrichthyes: Elasmobranchii) Abstract
AES 19th Annual Meeting, Manaus, Brazil
LITTLE, C.D. & BEMIS, W.E.

We present new illustrations and descriptions of the skeleton of the heterocercal tail of twelve species of sharks represented by multiple adult specimens. Nine species from the family Carcharhinidae were examined (bull shark, *Carcharhinus leucas*; blacktip shark, *C. limbatus*; dusky shark, *C. obscurus*; sandbar shark, *C. plumbeus*; tiger shark, *Galeocerdo cuvieri*; Atlantic sharpnose shark, *Rhizoprionodon terraenovae*; bonnethead, *Sphyrna tiburo*; great hammerhead, *S. mokarran*; and scalloped hammerhead, *S. lewini*). We also studied one species of Alopiidae (common thresher shark, *Alopias vulpinus*), as well as one species of Ginglymostomatidae (nurse shark, *Ginglymostoma cirratum*) and one species of Triakidae (smooth dogfish, *Mustelus canis*). Our most interesting observations concern anatomical relationships of the hemal arches and hypural elements that support the ventral fin-web of the hypochordal lobe of the caudal fin and the modified neural arches and spines that support the epichordal portion of the caudal fin. The patterns of these skeletal elements differ in many details from the patterns described previously for the heterocercal caudal fin of actinopterygians such as paddlefishes, and these differences offer insight into general aspects of the anatomy of heterocercal caudal fins. This paper is in honor of Hans-Peter Schultze and his many contributions to the anatomy and systematics of vertebrates. Supported by DEB-0075460 and the Jane H. Bemis Fund for Research in Natural History.

As in the case of the previous two abstracts, we are unable to publically access Little & Bemis' presentation in its entirety. However, the subject of the heterocercal tail is in itself a commanding contemplation: what is the purpose of this asymmetrical shape? Is there a meaning behind the Bull Shark's precise caudal configuration, as it relates to kinematics—to the forces involved in marine locomotion? I am as of yet unable to discover the truth in detail; but I can provide the basics.

In all sharks with the classic design, the heterocercal tail acts as a primitive but marvelous 'jet-propulsion' system. Every beat of the tail creates two small 'vortex rings' in the water; one ring spins clockwise, while the other spins counter-clockwise. As the rings spin together, they create between them a powerful 'jet flow', which is expelled behind the shark. Thus, as the shark continuously beats its heterocercal tail, it continuously generates vortex rings, which in turn maintain an unbroken and powerful jet stream. As the jet rapidly pulls water between the rings (and behind the shark), it simultaneously creates an equal and opposite 'reaction force', which pushes the shark forward. Much like in airplanes, this force can be thought of as 'forward thrust'.

It is incredible to realize that the shape, movement, and strength of the vortices required for thrust is in direct relation to the shape, movement, and strength of the shark's tail. From this relationship, we can make a miraculous deduction: that each shark species' unique tail shape will generate a unique vortex pattern. Every shark species thus creates a distinctive jet stream for locomotion, and thereby possesses a characteristic style of movement. As such, the term *Carcharhinus leucas* applies to not only a physical form, but also a school of motion. Poetically, this can be thought of as a kind of kata, consisting of vortices, jet streams, and thrust unique to the *Carcharhinus leucas* style.

2003

Occurrence of Two Species of Elasmobranchs, Carcharinus leucas and Pristis microdon, in Betsiboka River, West Madagascar

Cybium, 27 (3): 237-241

TANIUCHI, T. & ISHIHARA, H. & TANAKA, S. & HYODO, S. & MURAKAMI, M. & SÉRET, B.

RÉSUMÉ. – Présence de deux espèces d'Élasmobranches, *Carcharhinus leucas* et *Pristis microdon*, dans la rivière Betsiboka, oust de Madagascar.

En septembre 2001, une mission a été effectuée dans l'ouest de Madagascar pour étudier les Élasmobranches d'eau douce. Sept specimens d'Elasmobranches ont été récoltés dans le basin de la rivière Betsiboka, près de Marovoay, ouest de Madagascar: 1 mâle et 2 femelles de requin-bouledogue, *Carcharhinus leucas*, et 1 mâle et 3 femelles du poisson-scie à large dents, *Pristis microdon*. Les analyses ont montré que l'eau des sites présumés de capture de *P. microdon* était douce, mais qu'elle était de qualité incertaine pour les sites de captures de *C. leucas*. L'analyse de la composition du serum sanguine suggère que les specimens de *P. microdon* ont été capturés en eau douce et ceux de *C. leucas* en eau saumâtre.

Although they are principally marine, elasmobranchs have been known to occur in freshwater. There are many definitions of freshwater as stated by Schwartz (1995) and freshwater elasmobranchs have been defined by Zorzi (1995) as those << sharks and rays that frequent rivers and lakes >>. Compagno and Cook (1995) pointed out that 10 genera, four families and approximately 43 species of sharks and rays penetrate freshwater environments well beyond the tidal reaches of river mouths.

In Madagascar, the Bull Shark is sometimes known as "Ankiho Beloa". This local variant is relatively common, and has brought death in a remarkable way: ichthyosarcotoxism. In November 1993, the southeastern Malagasy city of Manakara witnessed a fatal mass poisoning, "attributed to the ingestion of the meat of a bull shark". The poison affected 188 inhabitants, and claimed about 30% of the total afflicted. Taniuchi et al. do not specify as to what may have led to the meat's toxicity; it is unclear whether that particular Bull Shark accumulated toxins from its environment (such as in the case of mercury poisoning) or suffered from a communicable disease. The flesh itself may have also naturally rotted postmortem, and been mistakenly ingested.

Whatever the cause, this extraordinary case illustrates a new, fascinating complication in the link between *Carcharhinus leucas* and *Homo sapiens*. Normally, each could consume the other; *Carcharhinus leucas* could build itself upon the flesh of man, and *Homo sapiens* could likewise build itself upon the flesh of shark. However, the Manakara poisoning represents a denial of this relationship; though this particular Bull Shark failed to protect itself from the fisheries of man, it nonetheless denied predatory victory by possessing corrupted, inaccessible flesh. There is, of course, no way that the Bull Shark could consciously will for this circumstance; however, it is fascinating to realize that the interspecies link broke as consequence of the shark's nature, not man's.

In addition to this fascinating anomaly, Tanuichi et al. also remark upon a curious observation made by the residents of Lake Kinkony: they report that some Ankiho Beloa, "never go downstream to the sea." This sounds suspiciously familiar; Lake Nicaragua likewise claimed an exclusively freshwater *Carcharhinus leucas*, but this was later falsified. Are the Ankiho Beloa of Lake Kinkony similarly misunderstood?

2004

Os Carcharhiniformes (Chondrichthyes, Neoselachii) da Bacia de Alvalade (Portugal)
Revista Española de Paleontología, 19 (1): 73-92
ANTUNES, M.T. & BALBINO, A.C.

The uppermost Miocene, Esbarrondadoiro Formation (Alvalade basin, Portugal) yielded more than 10 thousand Selachian teeth at Santa Margarida, Esbarrondadoiro and Vale de Zebro outcrops. Forty-five taxa were identified belonging to the orders Hexanchiformes, Squaliformes, Lamniformes, Carcharhiniformes, Torpediniformes and Myliobatiformes.

The Carcharhiniformes make up about 40% of the selachian fauna that has been identified in the studied area. The families Scyliorhinidae, Triakidae, Hemigaleidae, Carcharhinidae and Sphyrnidae, and fifteen species are recognized.

The distribution of the Carcharhiniformes by the Santa Margarida, Esbarrondadoiro and Vale de Zebro localities is shown (Table 1). The quantitative taxa distribution among these localities is very unequal. The predominant forms are *Premontreia (Oxyscyllium)* cf. *dachiardi*, *Mustelus* sp., *Paragaleus antunesi*, and especially *Carcharhinus* cf. *perezi*.

The different distribution of the Carcharhiniormes (as well as that of the other orders) by the three sites points out to distinct environments in the corresponding areas: Esbarrondadoiro indicates relatively deeper, rather still waters; Santa Margarida represents a very littoral area and rough waters; while Vale de Zebro was a (probably inner) part of a gulf with muddy bottoms.

Much of this territory is familiar, as we have encountered Antunes et al.'s Miocene fossil research on the Alvalade Basin before. This report is distinguished by new morphological comparisons between the fossil teeth of the 'Ancient Bull', *Carcharhinus* cf. *leucas*, and the modern teeth of four *Carcharhinus*: the Dusky Shark (*C. obscurus*), the Sandbar Shark (*C. plumbeus*), the Bignose Shark (*C. altimus*) and the contemporary *Carcharhinus leucas*. The four modern species were chosen because they shared broadly triangular upper teeth. The distinctions between them are slight, but enough to verify that *Carcharhinus* cf. *leucas* matches most closely in morphology to the modern Bull Shark.

Dusky Shark teeth have wider, thicker roots than those of the Miocene Bull, and a shorter labial (i.e., 'front' of the tooth, facing outwards towards the lips) face. Sandbar Shark teeth are more labio-lingually flattened than Bull Shark teeth, while Bignose Shark teeth are conversely taller and more tapered. *Carcharhinus* cf. *leucas* was once attributed to the fossil species *C. egertoni*, but incorrectly so; the apices of Miocene Bull Shark teeth are more distinctly tapered than those of *C. egertoni*.

Fourteen *Carcharhinus* cf. *leucas* teeth were referenced in this comparison: ten from Esbarrondadoiro and four from Vale de Zebro. As asserted in the abstract, the habitat types associated with these two locales were, "deeper, rather still waters" and a, "gulf with muddy bottoms". The abundance of *Carcharhininus* in the Alvalade Basin, combined with their relative absence off modern Portugal, indicates that this Miocene ecosystem benefited from a warmer marine environment than that of today. Antunes & Balbino also note that the relative absence of larger pelagic forms, "points out to a quite narrow gulf and not to an open Atlantic front." It is fascinating to consider these two elements as 'ingredients' for an environment conducive to *Carcharhinus* abundance.

2004

Estimates of Sharp Species Composition and Numbers
Associated with the Shark Fin Trade Based on
Hong Kong Auction Data
Journal of Northwest Atlantic Fishery Science, 35; 453-465
CLARKE, S.C. & MCALLISTER, M.K. & MICHIELSENS,
C.G.J.

The species composition and number of sharks used by the shark fin trade were estimated from a partial set of saily auction records for the world's largest shark fin trading centre in Hong Kong for the period October 1999 to March 2001. More than 10 000 lot descriptions of shark type, fin position, fin size and fin weight were translated and statistically modeled using Bayesian Markov Chain Monte Carlo methods (WinBUGS). These methods allowed a robust estimation of missing information in individual auction records, as well as of entire auctions for which no data are available, through a hierarchical model with uninformative priors. The model provides estimates of the complete data set for the sampled period, including the total auctioned weights of fins by shark type and fin position. Separate studies, undertaken in Hong Kong to genetically map trade names to species names, are being used to align the estimates with particular taxa. This paper demonstrates how the traded quantity estimates can be converted to the weight and number of sharks represented based on preliminary conversion factors from the literature and from this research. A potentially more robust Bayesian conversion algorithm, involving fin size-classes and stochastic relationships between fin lengths and fin weights, is outlined for future implementation.

Hong Kong is rumored to control half of the world's shark fin trade (estimates range from 50% to 85%), and is certainly anchored at the market's global center. At least 85 countries export unprocessed shark fins to Hong Kong; many of these are then re-exported to Mainland China for low-cost processing (rendering Hong Kong as a Chinese entrepôt). Clarke et al. suggest that collecting data at the Hongkonger-Chinese exchange, "allows fins to be characterized by shark type, fin position, and fin size." Given Hong Kong's monumental position in the shark fin market, this data could be, "extrapolated, with appropriate caveats, to depict the global trade."

Carcharhinus leucas is believed to share the trade name of Sha Qing with *C. amboinensis* (which is understandable, given the two species' immensely similar appearance). According to Clarke et al. (who collected auction data from October 1999 to March 2001), Sha Qing is estimated to compose 3.48% of the overall traded shark fin weight total. This figure translates to 40,924 kilograms in mean traded weight annually, with the following distributions by fin type: 5,052kg for dorsal fins; 4,012kg for caudal fins; 11,910kg for pectoral fins; and 19,936kg for unidentified fins. As the authors could neither observe every fin auction nor access every trade house's information, their assessment is far from complete; Clarke et al. suggest that their estimates represent about 20% of the total number of shark fins traded through Hong Kong, and 10% of the global market (that is, if Hong Kong controls half of the world's shark fin trade, as alleged).

Though not necessarily complete, this incredible wealth of information is an important complement to traditional catch and landing data, which has been "short of addressing the question of whether vulnerable shark species are being overexploited." Matching catch data with fin trade data is an excellent way to reduce such uncertainty.

2004

Age and Growth of the Bull Shark, Carcharhinus leucas,
from Southern Gulf of Mexico
Journal of Northwest Atlantic Fishery Science, 35: 367-374
CRUZ-MARTÍNEZ, A. & CHIAPPA-CARRARA, X. &
ARENAS-FUENTES, V.

Age and growth of bull shark, *Carcharhinus leucas*, was
investigated in the southern Gulf of Mexico (Veracruz and Campeche,
Mexico) from December 1993 through June 1997. Ninety-five specimens
were obtained from commercial fishery catches, and vertebrae were
examined from 20 males, 61 females and 14 individuals unidentified to sex.
Vertebrae were examined using five different techniques to enhance the
visibility of growth rings: i) alizarin red stain, ii) crystal violet stain, iii) X-
ray, iv) silver nitrate stain, and v) without staining. Verification of temporal
growth ring formation was done by the indirect method of marginal
increment analysis. An isometric relationship was found between growth
and length of centrum, is described by a linear equation. Age-at-maturity
was 10 years (204 cm total length, TL) for females and 9-10 years (190-200
cm TL) for males. The oldest female was 28 (256.0 cm TL), and the oldest
male was 23 (243.0 cm TL). The von Bertalanffy growth parameters were
estimated for the species (L_∞ = 256.4 cm TL, k = 0.1397 per year and t_o = -
1.935), for males (L_∞ = 248.4 cm TL, k = 0.1692 per year and t_o = -1.03),
and for females (L_∞ = 262.1 cm TL, k = 0.1235 per year and t_o = -2.44).
Sexual differences for each particular growth curve were found, L_∞ being
the parameter that showed the greatest difference between males and
females; females attain a larger size.

164

The epitome of growth for *Carcharhinus leucas* lies within the vertebrae: much like in trees, the growth rings of each vertebral centrum can be counted to determine age. Amazingly, each ring neatly corresponds to a year; the ring bands form in a seasonal pattern (perhaps in response to seasonal changes in diet and temperature), with calcium band deposition most likely taking place in the fall.

Cruz-Martínez et al. piece together a fascinating life history for the southern Gulf of Mexico's Bull Shark population. Young Bulls experience a high growth rate until sexual maturity at about 9 or 10 years of age. Females grow more slowly than males, but attain a larger size; furthermore, the females of the southern Gulf of Mexico may reach an older age than the males, as the sampling's oldest female was 28, while the oldest male was 23. Overall, *Carcharhinus leucas* grows relatively slowly, with a k estimate (or 'growth rate', from the von Bertalanffy growth model) of 0.12 per year for females, and 0.16 per year for males. This trait, in combination with the Bull Shark's late maturity, is characteristic of a K-selected species. Such organisms do not fare well in the face of overharvesting; the K-strategy is one of slow growth, large size, and late reproduction.

Cruz-Martínez et al. compare their results with the projections of other authors, and there is a shocking degree of variation. For example, Castro (1983), Compagno (1984), and Rodriguez de la Cruz (1996) all state that *Carcharhinus leucas* can attain a maximum size well beyond 300 cm (with the highest value being Castro's 350 cm). With a mean maximum length (L_∞) of 256 cm for the species (L_∞= 262 cm for females, L_∞ = 248 cm for males), Cruz-Martínez et al.'s results potentially challenge the accuracy of older studies. Conversely, they may also reflect a regional population difference; Bull Sharks from the southern Gulf of Mexico may simply be smaller than the global average.

2004

Catch Rates and Movements of Bull and Great
Hammerhead Sharks in the Lower Florida Keys Abstract
American Elasmobranch Society 20th Annual Meeting,
Norman, Oklahoma
HEITHAUS, M.R. & HUETER, R.E.

Bull sharks (*Carcharhinus leucas*) and hammerhead sharks (*Sphyrna mokarran*) are top predators in coastal waters of the Caribbean and the Gulf of Mexico, but little is known about their habitat use patterns or movements. We conducted a preliminary study of the distribution and movements of these sharks in near-shore habitats of the lower Florida Keys using catch rates, fixed-location acoustic monitoring systems and satellite telemetry. Catch rates of both bull and hammerhead sharks were influenced by habitat type and bait type. Despite being sympatric in many habitats, the few individuals monitored with acoustic telemetry and satellite telemetry as well as recaptures by fishermen suggest markedly different movements over the period of days. Generally, bull sharks made more localized movements over a period of weeks while four hammerhead sharks equipped with satellite transmitters moved several hundred kilometers east in a matter of days, and one shark made a dive to at least 300 meters depth. Bull sharks however, spent the majority of their time in much shallower water.

To juxtapose *Carcharhinus leucas* with *Sphyrna mokarran* is to enrich our appreciation of the species' ecological identity. Only through comparison is it most apparent that *C. leucas* has a kind of 'ecological edge' in the form of its preference for localized movements in shallow sunlit waters. Within the edge, it overlaps with *Sphyrna mokarran*, and in that sense shares a part of its experience and evolution with the influence of the Great Hammerhead. Beyond the edge, the Great Hammerhead pursues experience independent from the Bull Shark's ecological influence, and in that sense, seizes a unique, unmatchable identity distinct from that of *Carcharhinus leucas*. Likewise, the Bull Shark escapes the 'shared shallows' into estuarine and freshwater habitat—environments into which the Great Hammerhead cannot follow. Thus, the Bull too asserts a unique, independent experience—a kind of 'environmental identity' unmatchable by the Great Hammerhead, or any other shark for that matter.

This is the essence of 'niche', and it is a multifaceted, powerful concept. Niche is reflected in physicality, in behavior, and in environment—all are one with the same identity. This is what I mean by my statements of, "*Carcharhinus leucas* is not only a name for a shark, but also for a swimming pattern, a feeding preference, etc." As every facet of niche—physical form, environment, and behavior—is directly tied to 'the single identity', each can thus serve as a 'clue to identity'. For example, consider if we have never seen a Bull Shark in the wild, and have never directly observed its habitat preference—can we still deduce where it lives? The answer is yes, because the shark's physical form *reflects* the ecological niche *reflects* the physical environment; isn't it interesting that Bull Shark eyes are rather small? Why else would they be so inefficient at absorbing light, unless there was an abundance of sunlight to begin with? In which habitat does such abundance lie? Shallows and Shark—*they are one*.

2004

Does Freshness Count? Movements and Habitat Use by Young Bull Sharks in the Caloosahatchee River Abstract
AES 20th Annual Meeting, Norman, Oklahoma
HEUPEL, M.R.

The use of freshwater habitats by bull sharks, *Carcharhinus leucas*, has been widely reported in the literature. This species is known to utilize habitats far up rivers and into large lakes. There has been little data, however, concerning how bull sharks use these habitats. This presentation will examine the use of the Caloosahatchee River as a nursery area for neonate and young-of-the-year *C. leucas*. To examine habitat use and home range size 18 *C. leucas* were captured in the Caloosahatchee River in August 2003, and fitted with acoustic transmitters. The movement patterns of all 18 individuals were monitored via a series of 20 acoustic hydrophones. Monitored sharks had a restricted home range (c. 4 km2) near the river mouth for the first two months of the study. No sharks were caught outside this region although data from surveys in previous years have shown that young *C. leucas* are commonly found farther upstream at this time of year. Rainfall levels during the summer of 2003 were higher than average, causing lower than normal salinities within the Caloosahatchee River study site. Salinity at the mouth of the river in the first two months of the study averaged 0.2 ppt. By October salinity levels at the mouth of the river had increases to 11.2 ppt and shark home ranges has increased to over 27 km2. In December of 2003 salinity levels at the mouth of the river were 16.8 ppt and shark home ranges size had expanded to the full extent of the monitoring array, over 69 km2.

We need to reexamine the relationship between Freshwater and the Bull Shark. Freshwater has, historically, upheld a tradition of benevolence towards the Bull Shark; it is, in our cultural image, practice that *C. leucas* be allowed access to passages deep within the continent, without limitation or issue, via Freshwater's goodwill. However, the Caloosahatchee presents an alternative to this traditional idealism; for the first time in our journey, we see Freshwater contest the ability of the Bull Shark to explore.

Heupel's research indicates that, miraculously, *Carcharhinus leucas* can be *limited* by salinity. The idea is almost astonishing; the very identity of the Bull Shark is near synonymous with Freshwater, and yet, Freshwater resists. The young Bull Sharks of the Caloosahatchee didn't even cross the freshwater salinity threshold of 0.05ppt, but instead remained at a salinity 'gate' of 0.2ppt. It was only when the salinity *increased* that the sharks expanded their range and (assumedly) gained upriver access.

This complicates the bond between Freshwater and Bull Shark: it appears to be conditional. Though a euryhaline identity is still true-to-form for this species, it is enlightening to note that *Carcharhinus leucas* is not universally applicable to all rivers at all times—that life is too marvelously complex to condone a 'one size fits all' mentality: not every Bull Shark enters Freshwater.

For the Caloosahatchee River specifically, its seems that the pups are either too young to handle the full experience, or too adjusted to a brackish lifestyle to even bother with further exploration. Whatever the reason, we do see Freshwater act in a challenging capacity for this species in the Caloosahatchee: it offers resistance, and in that sense, provides definition. The Caloosahatchee Bulls are limited to a brackish-estuarine lifestyle in their first years; their diet, behavior, and physical composure are thus locally defined.

2004

Chapter 14: Elasmobranch Nutrition, Food Handling, and Feeding Techniques
In: The Elasmobranch Husbandry Manual: Captive Care of Sharks, Rays and their Relatives. Special Publication of the Ohio Biological Survey: 183-200
JANSE, M. & FIRCHAU, B. & MOHAN, P.J.

The correct nutrition of elasmobranchs in captivity is of fundamental importance to their health and survival. The optimal diet for a captive elasmobranch is a copy of its diet in the wild, both in quantity and quality. These parameters change seasonally, within species and between species. It is incumbent upon the aquarist to determine the appropriate diet for a given age class and species of elasmobranch. Feeding of elasmobranchs in captivity is mostly done with pre-frozen food, to eliminate the possibility of parasitic infection and to ensure the continuity of food availability. Loss of vitamins and minerals due to food transport, storage, and preparation make vitamin and mineral supplementation necessary. Of equal importance to nutritional content and food quantity, is the manner in which the food is stored and thawed, its hygienic preparation, and the techniques used for delivering the food to specific elasmobranchs. Target-feeding individual specimens is preferred, as this allows an accurate assessment of the status of each animal and facilitates the administration of medications. As a last resort, tube-feeding may be employed to prevent dehydration and excess loss of body weight in inappetant elasmobranchs.

This luminous caretaking manual is dazzlingly detailed, but offers very little specific data on *Carcharhinus leucas* nutritional requirements. The only key piece of information presented is a note on feeding rations: for juvenile Bull Sharks, the recommended portion is 3.4% of their body weight (BW) per week. Adults require less, sufficing with a decreased portion of 2.7% BW per week. Both suggestions originate from a 1994 study by Schmid & Murru, which concerned captive juvenile, subadult, and adult Bull Sharks kept in a warm aquarium with a temperature range of 23-25°C.

For Carcharhinids in general, a diet of large and small teleosts, crustaceans, cephalopods, and elasmobranchs is recommended. Janse et al. provide an excellent list of specific food items (most to the species level) and their essential nutritional information (which includes water content, protein, fat, and energy content). Amongst teleosts, Atlantic Herring (*Clupea harengus*) appear to pack the most nutritional punch, with a maximum energy content of 233 kcal/100g. Sprat (*Sprattus sprattus*), Atlantic Mackerel (*Scomber scombrus*), and Blue Runner (*Caranyx chrysos*) also boast high nutritional numbers (with a specific energy content range of 163-217 kcal/100g). Krill (*Euphausia superba*) is reported to be the most nutritional crustacean (91 kcal/100g). Squid of the genus *Loligo* are the cephalopod energy superlative (85 kcal/100g).

As in the case with all elasmobranchs, newly pupped Carcharhinids must first absorb their entire yolk before being fed in an aquarium setting. Once the last of their yolk has been absorbed, pups should be introduced to a wide variety of food types to encourage active feeding behavior. This is the only phase in captivity where live food is preferable; as the shark ages, it should slowly be weaned off live food and adjusted to nutritionally enhanced dead food (to reduce the risk of parasitism or infections).

2004

Osmoregulation and the Sodium Pump in the Bull Shark, Carcharhinus leucas from the Brisbane River, Queensland
American Elasmobranch Society 20th Annual Meeting, Norman, Oklahoma

MEISCHKE, L.J. & CUTLER, C. & HAZON, N. & CRAMB, G.

The Bull Shark, *Carcharhinus leucas*, is rare amongst the elasmobranchs in that it is able to tolerate both seawater and freshwater environments. The body fluids of *C. leucas* can range between 600 and 1100 mom kg-1, but are always hyperosmotic to either the seawater (around 900-1000 mOsm kg-1) or freshwater (less than 100 mOsm kg-1) environment in which they are acclimated. Despite this variability in osmotic concentration, the plasma salt concentrations are maintained at levels similar to that of marine teleosts. This is achieved by efficient transport and regulation of water and ions which requires the expression and function of a number of membrane transporters which are fundamental to osmoregulation, including the Na, K-ATPase (sodium pump). Using RT-PCR and 5'- and 3' RACE techniques we have amplified, cloned and sequenced cDNAs encoding the bull shark Na, K-ATPase alpha and beta subunits. The nucleotide and putative amino acid sequences have been analysed and compared to those published for other species including other stenohaline and euryhaline elasmobranchs. Further immunohistochemical and molecular studies are presently being conducted to characterize the expression and distribution of Na, K-ATPase alpha and beta subunits within the osmoregulatory tissues of bull sharks acclimated to FW or SW environments.

The infamous "sodium pump" must be discussed, if we are to fully appreciate the chemical magnificence and environmental mastery of *C. leucas*. I use the term "infamous" because this very pump is one of Biology 101's many classic intellectual thorns; it ranks with oxidation-reduction and the Krebs cycle as the most burdensome basic to ever be bequeathed. However, it is the mechanism essential to the Bull Shark's survival in varying salinities—a mechanism essential to our own survival, as animals living in balance, with a healthy concentration of water in our cells.

Our bodies cannot have too much water, and likewise, cannot live with too little; we must find a balanced concentration, so that our cells may neither shrivel nor explode. Water leaves or enters a cell in response to the cell's concentration of solutes, such as salt (sodium chloride, $NaCl$), sodium (Na), or potassium (K). If the concentration of solutes is higher inside a cell than outside, then water moves in (cell expansion); if the concentration of solutes is lower inside the cell than outside, then water moves out (cell shrinkage). In order to keep the cell at a healthy balance, the sodium pump actively alters the cellular concentration of solutes; specifically, the pump removes sodium (Na) from the cell, and simultaneously adds potassium (K). This process reduces the cellular solute concentration (of Na) and, by extension, reduces the risk of water rushing into the cell.

Without the sodium pump (whose proper name is Na, K-ATPase), water would rush into a cell unchecked, and cause it to explode. This structure is a core component of the grander process of osmoregulation, and a keystone to the Bull Shark's survival in both saltwater (which has a higher environmental concentration of solutes) and freshwater (which has a lower environmental concentration of solutes). The activity of sodium pumps generally increases in freshwater, to prevent cell explosion (hypotonicity).

2004

Fish Fauna of the Fitzroy River in the Kimberley Region of Western Australia—including the Bunuba, Gooniyandi, Ngarinyin, Nyikina and Walmajarri Aboriginal Names
Records of the Western Australian Museum, 22: 147-161
MORGAN, D.L. & ALLEN, M.G. & BEDFORD, P. & HORSTMAN, M.

This project surveyed the fish fauna of the Fitzroy River, one of Australia's largest river systems that remains unregulated, located in the Kimberley region of WA. A total of 37 fish species were recorded in the 70 sites sampled. Twenty-three of these species are freshwater fishes (i.e. they complete their life-cycle in freshwater), the remainder being of estuarine or marine origin that may spend part of their life-cycle in freshwater. The number of freshwater species in the Fitzroy River is high by Australian standards. Three of the freshwater fish species recorded are currently undescribed, and two have no formal common or scientific names, but do have Aboriginal names. Where possible, the English (common), scientific and Aboriginal names for the different species of the river are given. This includes the Aboriginal names of the fish for the following five languages (Bunuba, Gooniyandi, Ngarinyin, Nyikina and Walmajarri) of the Fitzroy River Valley. The fish fauna of the river was shown to be significantly different between each of the lower, middle and upper reaches of the main channel. Furthermore, the smaller tributaries and the upper gorge country sites were significantly different to those in the main channel, while the major billabongs of the river had fish assemblages significantly different to all sites with the exception of the middle reaches of the river.

The Dreamtime. Before empire. Before English. The way-back before. The proceeding after. The time beyond span. The place before birth. The being beyond death. The loop of creation. The names of entities present but not before. Born from ancient spirits, but not gods. Crawling, walking, shaping the landscape as they created. Rivers, gorges, billabongs, all formed by hands and tails. Of walkers. Consistent, but untouchable. Beyond and yet amongst. Corporeal in ourselves and our nature, and yet tangible when we break apart. Shells that form and crack. Teeth that bear and fall.

Bull Shark is Ngawoonkoo is Ngangu is *Carcharhinus leucas*. The namer is irellevant. The names are paper. The being is corporeal. The audience is the namer. The outside is the rider. Our hands are the maker. The shark is the placement. Ngawoonkoo is king beyond us. A river spirit. A carver in itself. It too connected. Broken into an unbroken loop. Ngangu breaks the surface. Its fin in the sun. Its shape in the river. Its path of the channel—the ancient highway carved by other Serpents. Sinusoidal crinkles and cruves. Breaking the landscape. Fresh to salt. Dark to light. Dawn to dusk. Dream to life. Breath to breathing. Jaw to meat. Meat to creation. Teeth emerge for a blink. Chocokablok with power. Breif. Ngawoonkoo moves quickly. Then disappears.

Ngangu can't be far. Our hands have placed it, within our reach. Paddle-push through the river. Dive down into sinking slow. Crocodile get Ngangu. Ngangu bite crocodile. Broken backs, unbreakable spirits, disincorporate, reincorporate. The Dreams are rewoven, and untearing. Ngawoonkoo swims into the Rainbow, and remerges as placement, for a moment. Ngawoonkoo is the river. The river mirrors the sky. The sky sits upon the land. The land births the trees. The trees kindle the fire. The fire tells the stories. It is how we share Ngawoonkoo. It is how Ngangu came to be.

2004

Checklist of the Helminth Parasites of the Vertebrates in Costa Rica

Revista de Biología Tropical, 52: 1-41

RODRIGUEZ-ORTIZ, B. & GARCIA-PRIETO, L. & PEREZ-PONCE DE LEON, G.

Helminth parasites of vertebrates have been studied in Costa Rica for more than 50 years. Survey work on this group of parasites is far from complete. We assembled a database with all the records of helminth parasites of wild and domestic vertebrates in Costa Rica. Information was obtained from different sources such as literature search (all published accounts) and parasite collections. Here we present a checklist with a parasite-host list as well as a host-parasite list. Up to now, 303 species have been recorded, including 81 species of digeneans, 23 monogeneans, 63 cestodes, 12 acanthocephalans, and 124 nematodes. In total, 108 species of vertebrates have been studied for helminthes in Costa Rica (31 species of fishes, 7 amphibians, 14 reptiles, 20 birds, and 36 mammals). This represents only 3.8% of the vertebrate fauna of Costa Rica since about 2,855 species of vertebrates occur in the country. Interestingly, 58 species (19.1%) were recorded as new species from Costa Rica and most of them are endemic to particular regions. Considering the valuable information that parasites provide because it is synergistic with all the information about the natural history of the hosts, helminth parasites of vertebrates in Costa Rica should be considered within any initiatives to accomplish the national inventory of biological resources.

The helminths are a group of paraphyletic parasitic worms, and fourteen have been recorded with *Carcharhinus leucas* in Costa Rica:

Class Cestoda:

Anthobothrium laciniatum

Callitetrarhynchus gracilis

Dasyrhynchus variouncinatus

Nybelinia bisulcata

Otobothrium penetrans

Cathetocephalus thatcheri

Phoreiobothrium triloculatum

Platybothrium hypoprioni

Phyllobothrium leuci

Phyllobothrium nicaraguensis

Phyllobothrium pristis

Class Monogenea:

Dermophthirius maccallumi

Erpocotyle carcharhini

Heteroncocotyle leucas

It is fitting to observe that some of these helminths bear (or, poetically, 'steal') the taxonomic name of their host (e.g., *Phyllobothrium leuci* and *Heteroncocotyle leucas*).

2004

Distribution of Immature Bull Sharks (Carcharhinus leucas) in a Southwest Florida Estuary Abstract
American Elasmobranch Society 20th Annual Meeting, Norman, Oklahoma
SIMPFENDORFER, C.A. & FREITAS, G.G. & WILEY, T.R. & HEUPEL, M.R.

The distribution and salinity preference of immature bull sharks (*Carcharhinus leucas*) was examined based on the results of longline surveys in three adjacent estuarine habitats in southwest Florida: the Caloosahatchee River, San Carlos Bay and Pine Island Sound. Size frequency distributions were significantly different between each of these areas indicating the occurrence of age-based habitat partitioning. Neonate and young-of-the-year animals occurred in the Caloosahatchee River and juveniles older than 1 year occurred in the adjacent embayments. The existence of habitat partitioning may reduce intra-specific predation risk, and so increase survival of young animals. The catch rates of juvenile *C. leucas* were highest in areas with salinities between 10 and 20 ppt. Thus although they are able to osmoregulate in salinities from fresh to fully marine, juvenile, *C. leucas* may, have a salinity preference. Reasons for this preference are unknown, but need to be further investigated.

The risk of intra-specific predation—essentially, cannibalism—is a concept new to us: apparently, *Carcharhinus leucas* consumes members of its own kind. Though Bull Shark cannibalism is casually mentioned here, the idea is worth some serious discussion: what are the implications of *Carcharhinus leucas* cannibalism—or cannibalism in general? Why has it evolved in certain species—including, in extreme cases, our own?

The act of cannibalism is almost supernatural, as it undermines the entire framework of species evolution, survival, and reproduction. A species evolves to survive and procreate—to 'make more of it', not less: for a species to 'devour itself' is to decrease its genetic diversity, and thereby risk its ability to successfully reproduce. Why would something so damaging to the gene pool be even *allowed* for evolution?

Could cannibalism be simply a reaction to extreme hunger? For many cases of human cannibalism, it is; however, certain groups (both past and present) have actively elected for cannibalism as an ultimate form of revenge, fantasy fulfillment, or sacred ritual—in such cases, hunger is not a key motive. For the Bull Shark, it is hard to determine whether the practice emerges from starvation, or for similarly nonessential reasons: eerily enough, a fellow *Carcharhinus leucas* could be 'just there' for the taking.

Motive aside, Bull Shark cannibalism in Florida is particularly disquieting, as it has possibly affected the local population's evolution. Simpfendorfer et al. suggest that neonate and young-of-the-year Bull Sharks use the Caloosahatchee to essentially 'hide' from their older kin. If this is true, then the 'pressure to hide'—the pervasiveness of cannibalism— must have been great. Larger *Carcharhinus leucas* may select against pups in San Carlos Bay and Pine Island Sound by simply eating them. Thus, the cannibal-free Caloosahatchee could have emerged as an evolutionary haven for younger Bull Sharks.

2004

Collections of the Museu Oceanográfico do Vale do Itajai
I. Catalog of Cartilaginous Fshes (Myxini,
Cephalaspidomorphi, Elasmobranchii, Holocephali)
Mare Magnum, 1 (2): 1-125
SOTO, J.M.R. & MINCARONE, M.M.

The type and non-type specimens of extant cartilaginous fishes
(hagfishes, lampreys, sharks, batoids, and chimaeras) collected through
2004 and catalogued in the collection of the Museu Oceanográfico do Vale
do Itajaí (MOVI) are listed. Included in these records are 4,823 specimens
in 1,538 lots representing 250 species. The MOVI collection of
cartilaginous fishes contains 7 holotypes and 48 paratypes of 9 species.
Most of the collection is composed of species from the Brazilian marine
fauna, especially those from the southern region; a few lots were collected
beyond Brazilian waters or are specimens donated by other institutions.
This catalog is organized as two lists: taxonomic list of species and list of
lots. The lists are arranged by class, order and family. Within families, taxa
are arranged alphabetically by genus and then species. Information for each
entry includes genus, species, author, year of publication, MOVI catalog
number, number of specimens, nature of the material collected, sex, size
range, location (ocean, country, state, county, coordinates, depth), vessel,
collection method, collector, collection date, donor, donation date,
identifier, and date of identification. Remarks pertaining to specimens
contained within a lot are also included when necessary.

At the time of Soto & Mincarone's publication, the Museu Oceanográfico do Vale do Itajaí possessed two specimens of *Carcharhinus leucas*. They are catalogued as:

MOVI 00073 (formerly MCNC) (1) jaws, f. mat. (2800 mm TL)
South Atlantic, Brazil, RS, Imbé, 30 km off Imbé, 25-31 m, FV "Laureano V", bottom gillnet J.M.R. Soto *leg.*, 15.i.1989, J.M.R. Soto *don.*, 15.i.1989, J.M.R. Soto *det.*, 14.vi.1991

MOVI 10179 (1) photographs, f. mat. (approx.. 2.3 m TL, 118 kg)
South Atlantic, Brazil, AM, Iranduba, Solimões river, 30 km from Manaus, 03°15'S, 60°20'W W. Damasceno *leg.*, 17.i.1994, J.M.R. Soto *don.*, 20.iv.1998, J.M.R. Soto *det.*, 20.iv.1998

Specimens like these are treasures: they capture the biodiversity of their original locales, and enrich the collective understanding of ichthyological systematics. Though each holding is valuable in its own right, type specimens—the essential 'archetypes' of a species—are particularly coveted. Type specimens anchor the scientific name: they are the essential physical examples of a particular species, and are sometimes the first individuals to ever be associated with a particular taxon. Perhaps the 'most prized' variety of type specimen is a holotype—the original physical specimen used to describe and establish a species. Though many holotypes exist for the Bull Shark's synonyms (such as SU 11890 for *Carcharhinus azureus*), none exist for the original *Carcharhinus leucas*; instead, there are two syntypes (MNHN A-9650 and MNHN A-9652).

2004

Elasmobranchs in the Fitzroy River, Western Australia
Report to the Natural Heritage Trust, 30pp.
THORBURN, D.C. & MORGAN, D.L. & ROWLAND, A.J. &
GILL, H.

In Australia, only two rays, *Pristis microdon* (freshwater sawfish) and *Himantura chaophraya* (freshwater whipray), and three sharks *Carcharhinus leucas* (bull shark), *Glyphis* sp. A (speartooth shark) and *Glyphis* sp. C (northern river shark), occur in oligohaline environments of the upper reaches of rivers far from the coast. While information on the occurrence, biology and distribution of these 'freshwater' species remains limited, recent surveys of rivers throughout northern Australia by Thorburn *et al*. (2003), and Morgan *et al*. (2002, in press) documenting the inland fishes of the Fitzroy River, have identified the Fitzroy River, Western Australia, as a significant habitat for several of these rare species.

Both Thorburn *et al*. (2003) and Morgan *et al*. (2002) recorded three of the above elasmobranch species in the Fitzroy River, these being *P. microdon*, *H. chaophraya* and *C. leucas*, with each being found in estuarine and freshwater sites throughout the river (Figure 1). Furthermore, a single *Glyphis* sp. C was collected from the nearby marine tidal creek, Doctors Creek (Figure 1). During a concurrently run study, a total of six *Glyphis* sp. C were also collected from Doctors Creek and an additional survey site south of Derby (Thorburn *et al*. 2004). The significance of the Fitzroy River elasmobranch assemblage is further illustrated by the high conservation value of these species (Table 1).

The Fitzroy River is an excellent model system for the study of freshwater elasmobranchs and their conservation requirements. The River supports four (possibly five, in the case of *Glyphis* sp. C) elasmobranch species, and presents each with a challenging mix of natural seasonalities and unnatural human development. During the wet season, the Fitzroy is rich and unbroken; *Carcharhinus leucas* and its kin penetrate deep into the basin via a flushed freshwater mainstem. As the dry season approaches, the flow of freshwater decreases, and the Fritzroy shrinks into a series of shallow pools. Tidal influence becomes the river mouth's dominant force of water movement, and subsequently increases the salinity of the Fitzroy's lower reaches.

During the dry season, the ability to travel upriver vastly diminishes. Elasmobranchs such as *Carcharhinus leucas* become concentrated within freshwater pools, and risk capture by human anglers. Furthermore, the Camballan Weir (located at the middle reach of the Fitzroy) permanently blocks *Carcharhinus leucas* and the others from entering the upper reach. This manmade obstacle is impassable during the dry season, and further increases the risk of predation by 'locking' the elasmobranchs below the Weir; most of the individuals recorded by Thorburn et al. were captured in a pool immediately downriver from the Weir (which is itself a popular fishing spot).

Elasmobranchs are inherently sensitive to overexploitation due to their notoriously slow growth and reproduction rates; however, the Fitzroy's freshwater species are especially vulnerable, thanks to the compounding risks of the dry season and the Camballan Weir. This challenging dynamic may not be unique to the Fitzroy, and provokes a powerful question regarding *Carcharhinus leucas* on the global scale: are Bull Sharks at a higher conservation risk in freshwater?

2004

Status of Freshwater and Estuarine Elasmobranchs in
Northern Australia
Final Report to Natural Heritage Trust: 1-75
THORBURN, D.C. & PEVERELL, S.C. & STEVENS, J.D. &
LAST, P.R. & ROWLAND, A.J.

Australia has an extremely rich shark and ray fauna with the most recent taxonomic review estimating that of the approximately 1025 species of sharks and rays worldwide (Leonard Compagno, South African Museum, Cape Town, pers. comm), at least 297 species inhabit Australian waters (Last and Stevens 1994). Of these species more than half (54%) are endemic to Australian waters (Shark Advisory Group 2001).

While there is still much to be learnt of the taxonomy, biology and distribution of chondrichthyan species found in Australia, a recent comprehensive review of the fauna (Last and Stevens 1994), and international synopses such as those produced by the Food and Agriculture Organisation of the United Nations (FAO, several authors, 1998; 1999), reveal significant advances in our knowledge compared to the earlier accounts of Whitley (1940), covering 162 species, and of Paxton *et al.* (1989), covering 177 species.

3.1.2 The estuarine and freshwater component

Some 118 species (40%) of Australian elasmobranchs penetrate estuaries and the lower freshwater reaches of rivers (Last 2002). However, only two rays (*Pristis microdon* and *Himantura chaophraya*) and three sharks (*Carcharhinus leucas*, *Glyphis* sp. A and *Glyphis* sp. C) occur in oligohaline environments of the upper reaches of rivers far from the coast.

Carcharhinus leucas appears to primarily use Northern Australia's river systems as a nursery: by and large, the individuals captured from Queensland, Northern Territory, and Western Australia were mostly 1m or less in total length. Larger Bull Sharks were captured as well, but the dominant size class (75%) was that of juveniles and young-of-the-year (which bore the telltale umbilical scar, located on the belly between the origin of the pectoral fins). Regardless of size, Bull Sharks were captured in an impressive variety of habitats, and demonstrated no preference. Wide ranges were observed not only for salinity (0-37.3ppt), but also for water clarity (5-273 cm secchi depth), temperature (22.1-32.5°C), flow rate (0-0.5ms^{-1}) and depth (>5-21m). Overall, *C. leucas* comprised 24% of the total catch (making it this study's most commonly encountered elasmobranch) and seemed to be particularly abundant in the Daly and Mission rivers.

Aggregations of *Carcharhinus leucas* were found below man-made obstructions like rock bars, road crossings, or barrages. According to the authors, these aggregations "might indicate that this species would travel much further inland if unimpeded." This is a serious ecological concern, as juvenile Bull Sharks utilize the upper reaches for food and protection from larger predators. With these man-made obstacles, *Carcharhinus leucas* is forced to congregate downriver—to accumulate in a spatially restricted habitat (especially during the dry season). This kind of accumulation forces the juveniles to adjust their natural regime, and remain closer than usual to brackish or estuarine habitat.

High numbers of *Carcharhinus leucas* were captured via gillnet, as the sharks would pursue entangled teleosts with rapacity. Cannibalism was also documented in the Daly River; Thourburn et al. make note of a particular specimen hooked on a hand line, which upon capture was, "pursued and attacked by other bull sharks."

2004

Multidirectional Movements of Sportfish Species Between an Estuarine No-Take Zone and Surrounding Waters of the Indian River Lagoon, Florida
Fishery Bulletin, 102 (3): 533-544

TREMAIN, D.M. & HARNDEN, C.W. & ADAMS, D.H.

We examined movement patterns of sportfish that were tagged in the northern Indian River Lagoon, Florida, between 1990 and 1999 to assess the degree of fish exchange between an estuarine no-take zone (NTZ) and surrounding waters. The tagged fish were from seven species: red drum (*Sciaenops ocellatus*); black drum (*Pogonias cromis*); sheepshead (*Archosargus probatocephalus*); common snook (*Centropomus undecimalis*); spotted seatrout (*Cynoscion nebulosus*); bull shark (*Carcharhinus leucas*); and crevalle jack (*Caranx hippos*). A total of 403 tagged fish were recaptured during the study period, including 65 individuals that emigrated from the NTZ and 16 individuals that immigrated into the NTZ from surrounding waters of the lagoon. Migration distances between the original tagging location and the sites were from 0 to 150 km, and these migration distances appeared to be influenced by the proximity of the NTZ to spawning areas or other habitats that are important to specific life-history stages of individual species. Fish that immigrated into the NTZ moved distances ranging from approximately 10 to 75 km. Recapture rates for sportfish species that migrated across the NTZ boundary suggested that more individuals may move into the protected habitats than move out. These data demonstrated that although this estuarine no-take reserve can protect species from fishing, it may also serve to extract exploitable individuals from surrounding fisheries.

This is our first encounter with a No Take Zone (NTZ), and *Carcharhinus leucas* plays a minimal role (only 25 individuals were tagged). With such a small sample, it is extremely difficult to determine the NTZ's effect on the Bull Shark's health; only one individual was actually tagged within the No Take Zone's boundaries, while the other 24 were captured somewhere in the Indian River Lagoon outside of the NTZ area. *Carcharhinus leucas* was furthermore never recaptured inside the NTZ, so there is no direct evidence of this species' immigration into the No Take Zone's protected waters.

Although we cannot securely determine the NTZ's affects on *C. leucas*, we can estimate its positive impact on other species. Tremain et al. document high site fidelity for red drum, black drum, and spotted seatrout—all species affirmed to derive a long-term benefit from the NTZ's protection. Indeed, the largest tagged individuals of each drum species were caught in the NTZ area, not the surrounding lagoon; this implies that the NTZ's lack of a fishing pressure may allow for each of these drum species to attain a larger size. Migratory fish like permit, common snook, and gray snapper may also derive a short-term benefit from the NTZ, as, "seasonal movements could bring them into contact with the protected habitats."

The ambition of a No Take Zone is to enrich adjacent fisheries by providing them with healthier, larger, more fecund individuals; a fish living in such protected waters may live longer and reproduce more often (or at the very least, attain a larger size). Ideally, this kind of individual would 'spillover' into unprotected habitat, and thereby 'boost' the health of the ecosystem and the commercial stock it supports. Unfortunately, this study by Tremain et al. remains quite limited in scope; as such, their assessment of the NTZ's positive ecological (and fisheries) impact is cautiously optimistic.

2005

Insight into Migration Patterns of Bull Sharks in the
South Pacific Abstract
American Elasmobranch Society 21st Annual Meeting,
Tampa, Florida
BRUNNSCHWEILER, J.M.

Data on habitat use and seasonal movements are essential for designing conservation strategies, yet such data are rarely available for large marine animals such as sharks. In this study we equipped eleven bull sharks, *Carcharhinus leucas*, from a Fijian population with pop-up satellite tags to test the hypothesis that bull sharks migrate into nursery grounds. Individual tags remained attached for two to seven months. The pop-up locations give insight into movement patterns and distribution of bull sharks in the South Pacific. They further underscore the need for international cooperation in devising conservation plans.

As you can see, we have absolutely no details regarding the habitat use, movement patterns, or conservation needs of Fijian *Carcharhinus leucas*. This is the unfortunate consequence of a brief AES abstract (though, to be fair, the entire data-laden presentations cannot themselves be archived, as they are purely oral). Nevertheless, Brunnschweiler identifies pop-up satellite tags (also known as pop-up satellite archival tags, or PSATs) as the instruments of his method—an exciting key detail, given the incredible data potential of these instructive gadgets.

Pop-up satellite archival tags are small (typically under 100 g) data loggers designed to track the movements of large migratory animals such as sea turtles, swordfish, and sharks. PSATs are photosensitive and can record temperature, pressure, position, oxygen levels, and even magnetics (some PSATs possess magnetometers designed to reference the Earth's magnetic field). Geolocation can be estimated via the combined efforts of the magnetometer (for latitude) and the photometer (for longitude). Some PSATs can, amazingly, identify their location via the photometer alone: latitude is determined by recording the length of day (from first light to last light), while longitude is determined by computing the 'noon time' from this estimate. PSATs typically record data in the span of several weeks or months. Although data is physically stored within the PSAT itself, it doesn't have to be physically recovered: a PSAT will detach from its host at a predetermined date, and rise to the surface to automatically transmit satellite data.

Though PSATs are clearly incredible as research tools, they have limitations. Many can be subject to biofouling, power loss, or even predation (ingested PSATs will immediately record a loss of light and stability in temperature). Pressure limits (up to 2000m) and financial costs (up to $4,000) are additionally compelling obstacles to PSAT employment.

2005

Distribution of Elasmobranchs in the Brazilian Amazon River Floodplain Abstract
AES 21ˢᵗ Annual Meeting, Tampa, Florida
CHARVET-ALMEIDA, P. & ALMEIDA, M.P. & ALBERNAZ,
A.L.M.

The Amazon River floodplain is associated with white water rivers and presents seasonal changes that are closely linked to the hydrologic cycle. The information available on the elasmobranch species and respective distribution in the Solimões- Amazonas River system was very limited. An expedition was organized in September and October 2003 to collect data on the distribution of elasmobranchs, among other species groups. Twenty-six points of sampling were spread along over 3,000 km in the Brazilian portion of the Amazon River floodplain. Daily bottom long-line fisheries (captures), direct observation of specimens (registers) and interviews (reports) were used to provide evidence on the local elasmobranch species. Frequency of occurrence and biomass were calculated for each species. Water parameters were also noted. The results correspond to a unique specific study on the diversity of elasmobranchs present in this aquatic system. Information related to three elasmobranch families were obtained, namely: Pristidae, Carcharhinidae and Potamotrygonidae. The species that presented a wider distribution were: *Paratrygon aiereba, Plesiotrygon iwamae*, *Potamotrygon motoro*, *Pristis perotteti* and *Carcharhinus leucas*. Some potamotrygonid species were only observed in certain areas.

The Amazon's elasmobranch pallet is intriguingly similar to that of Northern Australia: both claim *Carcharhinus leucas*, sawfish, and a handful of freshwater rays. This particular combination may be a staple of a freshwater elasmobranch system; it would be interesting to explore whether or not the presence of one species (such as *Carcharhinus leucas*) directly determines the distribution of others (e.g., would the Bull Shark's midwater niche interfere with the survival of other large elasmobranchs, but allow for the existence of benthic hunters such as the freshwater sawfish and rays?)

In any case, it is wondrous to consider *Carcharhinus leucas* as a part of the world's most biodiverse freshwater ecosystem. Amazonian Bull Sharks may share the same waters as the mighty arapaima, the eerie pacu, or the infamous piranha—the same waters plied by caiman, anaconda, and jaguar. Though we cannot know the details from Charvet-Almeida et al.'s oral presentation, we do know that *Carcharhinus leucas* has been captured as far upriver as Iquitos, Peru (4,000km from the sea). Given this information, it is likely that *C. leucas* has access to the entire floodplain.

With such high biodiversity (the number of fish species alone totals to be about 5,600), the Amazon basin has nurtured an unsurprisingly rich variety of animal myths and legends. Colorful tales such as that of the seductive Bufeo Colorado (a river dolphin disguised as an attractive fisherman bent on courting women) or the mighty Pirarucu (a former warrior transformed by the gods into an arapaima, as a punishment for his vanity and pride) leave us to wonder: does the Bull Shark have its own legend in the Amazon?

There is only one shark story from the southern Americas that I know of, but it is very removed from the Amazonian basin: Cipactli—a primeval sea monster with an insatiable appetite, and one of the original forces behind the creation of the world.

2005

World Record Sized Giant Bull Shark Caught at Chennai Coast
CMFRI Newsletter, 107: 5
CMFRI

World record sized giant bull shark caught at Chennai coast

A giant sized female bull shark *Carcharhinus leucas* measuring 356 cm in total length with a weight of 320 kg caught by a mechanized gillnet, was landed on 22-6-2005 at Kasimedu Fisheries Harbour, Chennai. The present record of bull shark *C. leucas* is the largest recorded from Indian Ocean as well as the largest bull shark ever recorded in the world. The shark was auctioned at the landing centre for Rs. 32,000/-.

Large size bull sharks are caught by the gill nets regularly and the gillnet fishermen get good revenue. (*Madras Research Centre*)

The CMFRI (Central Marine Fisheries Research Institute) is an extensive research network concerning the biology and conservation of India's tropical marine fisheries resources. Headquarted at Kochi, the CMFRI boasts seven research centers (Mumbai, Karwar, Mangalore, Kozhikode, Vizhinjam, Tuticorin, and Chennai) as well as 18 field and regional offices. Chennai is home to the Madras Research Centre, which supports a rich diversity of projects (coral reef restoration, lobster mariculture and elasmobranch resource assessment are just some of the Centre's many colorful research foci).

The enormous Chennai Bull Shark is a massively important confirmation of the maximum size: a whopping 320 kg and 356 cm in total length meets (and, in many cases, exceeds) the projected expectations of Castro (1983), Compagno (1984), Bianchi (1986), Rodriguez de la Cruz (1996), and Cruz-Martínez (2004). Apart from being an impressive superlative, this maximum is a very valuable piece of information, as it elucidates our understanding of the Bull Shark's developmental potential. Now that we know that *C. leucas* can exceed 300 kg, we can begin to assess a new dimension of the species' ecological health: theoretically, the presence of such large individuals may indicate a healthy ecosystem fit for supporting superlative *C. leucas*. Likewise, the absence of such large individuals may be a sign of the ecosystem's failing health and inability to provide suitable habitat for Bull Sharks (or other large predators).

Of course, such assessments are preliminary, and as of now, imprecise. However, the capture of a 320 kg Bull Shark gives us hope: that in the face of increasing overexploitation and human development, *Carcharhinus leucas* can still, in some areas, thrive. The day we are to lose our massive Bulls– the day we no longer capture 300+ kg individuals – will portend the species' jeopardy; for now, we seem to be relatively safe.

2005

Checklist of Philippine Chondrichthyes
Report CSIRO Marine Laboratories, 243: 1-109
COMPAGNO, L.J.V. & LAST, P.R. & STEVENS, J.D. &
ALAVA, M.N.R.

Since the first publication on Philippines fishes in 1706, naturalist and ichthyologists have attempted to define and describe the diversity of this rich and biogeographically important fauna. The emphasis has been on fishes generally but these studies have also contributed greatly to our knowledge of chondrichthyans in the region, as well as across the broader Indo-West Pacific. An annotated checklist of cartilaginous fishes of the Philippines is compiled based on historical information and new data. A Taiwanese deepwater trawl survey off Luzon in 1995 produced specimens of 15 species including 12 new records for the Philippines and a few species new to science. Soon after, a major survey of fish markets in the southern Philippines, funded by the World Wildlife Fund, resulted in the collection and storage of specimens of 54 species, of which 41 were new records for the Philippines. Approximately 164 species representing 44 families and 83 genera of cartilaginous fishes have been recorded from Philippine seas from all available sources. The checklist includes 129 valid species records that are based on specimens and unambiguous literature accounts, and 35 additional doubtful records that are mostly based on literature. Of the valid species, about 111 were wholly or partly based on literature records; 109 species wholly or partly on specimens from all sources including museum collections. At least 24 species, mostly collected during the recent surveys, are new to science.

The Western Indo-Pacific is largely unexplored, yet home to the, "richest diversity of marine species in the world". According to Compagno et al., it may even boast the world's highest chondrichthyan biodiversity, as the list of newly discovered species seems to grow exponentially. However, the region is plagued with neglect and overexploitation, as it claims, "some of the most heavily exploited sharks and rays and amongst the least well known species in terms of their resource size and biology."

The Philippines is of particular concern, as its many colorful chondrichthyans severely lack both data and appropriate fisheries management. Manila was once a stronghold for research on Western Indo-Pacific chondrichthyans, as it boasted, "one of the most complete ichthyological libraries in the world at the Philippine Bureau of Sciences". However, the entirety of this collection—including the Bureau library and laboratories—was completely destroyed by the Japanese during World War II. Since then, the Philippines suffered from an extensive gap in knowledge on the systematics, biology, and conservation needs of its local sharks, rays, and chimaeras.

Thanks to an international rejuvenation in shark taxonomic research during the 1950s and 1960s, some interest has been renewed to the area. However, it wasn't until 1998 that the World Wildlife Fund (WWF) spearheaded the first major effort to identify Philippine chondrichtyhans (particularly from the Sulu Sea—a biodiversity hotspot) and begin their conservation assessment. This effort—the Elasmobranch Biology and Conservation Project—essentially, "established a data baseline of sharks, rays and chimaeras currently existing in the Sulu Sea and nearby areas". It is a core component of Compagno et al.'s current checklist, and provider of three records for *Carcharhinus leucas* (curiously, two make note of the first dorsal fin's unusually posterior placement).

2005

Distribution and Movements of Juvenile Bull Sharks,
Carcharhinus leucas, in the Indian River Lagoon System,
Florida Abstract
American Elasmobranch Society 21st Annual Meeting,
Tampa, Florida
CURTIS, T.H.

The Indian River Lagoon system along Florida's Atlantic coast is a nursery ground for bull sharks (Carcharhinus leucas). Since bull sharks are a component of Florida's shark fisheries, proper management requires a better understanding of their ecology within their vital nursery areas. A sampling program utilizing longlines and rod and reel has been initiated to estimate the current abundance and distribution of bull sharks in this estuary. Tagging and acoustic telemetry are being used to investigate the movements and habitat use of the young sharks. To date, sampling efforts have yielded the capture of 20 young-of-the-year and juvenile bull sharks (54-94 cm FL). They were captured over a broad range of salinities, depths, and oxygen concentrations, and only in temperatures > 20°C. Four sharks have been actively tracked, providing over 65 hours of movement data. Based on these preliminary results, the daily movements of these sharks appear to be confined to comparatively small core use areas (p< 4 km²). There were no obvious changes in movement patterns between day and night. Continued tagging and tracking efforts will provide a clearer understanding of how this important predator utilizes its nursery habitats.

Between the nurseries of the Indian River Lagoon and the Caloosahatchee River, the U.S. State of Florida may be safely nicknamed the 'Eden of American Bull Sharks'. To confirm a nursery in the Indian River Lagoon is especially exciting, as we are by now quite familiar with this area and its importance to fisheries management. The potential benefits of the Indian River Lagoon No Take Zone (NTZ), aforementioned by Tremain et al. (2004), may spillover to the local populace of neonate *Carcharhinus leucas*—either directly via the NTZ's physical protection, or indirectly as a result of the NTZ's potential to enrich and replenish adjacent stocks of Bull Shark prey items (such as red drum).

The protection offered by an NTZ may be particularly significant in relation to the neonate 'core use areas' as defined by Curtis. With a spatial maximum of only 4km^2, such areas could be relatively easy to delineate for the purpose of establishing a *Carcharhinus leucas* No Take Zone. Such an NTZ may have an enormously positive impact on the health and wellbeing of neonate Bull Sharks, and would doubly serve as an exceptional model system for the study of juvenile habitat use and population dynamics. Unprotected Bull Shark core use areas could still be a prime research focus: the concentration of spatially limited neonates makes for an excellent research precedent, and could maximally benefit the surveys of continued tagging and monitoring efforts.

Curtis' note on the 20°C temperature minimum is also intriguing: could this be a baseline for the species' thermal tolerance? It seems that little else determines the Bull's habitat use or migration patterns other than temperature; as reaffirmed by Curtis, *C. leucas* has yet to demonstrate a tangible preference in salinity, depth, or oxygen concentration. Given that the Bull Shark is typically characterized as a tropical species, a temperature threshold of 20°C appears to be a fitting minimum.

2005

Geographical and Temporal Variation in Length Distributions of Six Species of Shark Taken in the Bottom Longline Fishery off the Southeastern United States
Abstract
American Elasmobranch Society 21st Annual Meeting, Tampa, Florida
FORD, T. & GODDARD, N. & SICELOFF, L. & MORGAN, A. & BURGESS, G.H.

For any species subject to commercial fishing, changes in size distribution over time may indicate larger population-wide trends such as over-fishing and compensation. Thus, accurate data on catch size are important for assessment of current fishery management strategies. Seasonal changes in size may also indicate sex or size specific aggregation or be reflective of migratory patterns. We present an annual and monthly analyses of fork length (FL) distributions by sex of six coastal commonly targeted shark species in the bottom longline fishery off the southeastern United States. Length-frequency for *Carcharhinus plumbeus*, *C. limbatus*, *C. leucas*, *Sphyrna lewini*, *S. mokarran*, and *Rhizoprionodon terraenovae* were collected by the Commercial Shark Fishery Observer Program (CSFOP) between 1994 and 2004. Data were analyzed for differences in FL for males and females based on the fishing region, month and year. Significant differences in length frequency distributions were determined over time and between geographic areas. Factors influencing seasonal and regional patterns as well as long-term shifts in size classes will be discussed.

The Commercial Shark Fishery Observation Program (CSFOP) focused primarily on the captures of commercial bottom longline vessels targeting large coastal sharks. Between its 1994 inception and 2005, CSFOP reported 57,265 sharks representing 34 different species. Below is the Percent Total Catch for each—an interesting glimpse into their fisheries abundance within the U.S. Southern Atlantic and Gulf of Mexico:

Sandbar Shark	34.15%	Bonnethead Shark	0.12%
Atlantic Sharpnose Shark	27.35%	Florida Smoothhound	0.06%
Tiger Shark	10.71%	Bignose Shark	0.05%
Blacktip Shark	9.55%	Caribbean Reef Shark	0.03%
Blacknose Shark	4.43%	White Shark	0.03%
Dusky Shark	2.61%	Shortfin Mako Shark	0.02%
Nurse Shark	2.51%	Sixgill Shark	0.02%
Scalloped Hammerhead	1.56%	Smooth Hammerhead	0.01%
Bull Shark	**1.25%**	Sevengill Shark	0.01%
Spinner Shark	1.15%	Common Thresher Shark	0.01%
Smooth Dogfish	1.11%	Bigeye Thresher Shark	0.01%
Silky Shark	0.94%	Spiny Dogfish	0.01%
Great Hammerhead	0.71%	Roughskin Spiny Dogfish	0.01%
Lemon Shark	0.60%	Galapagos Shark	0.00%
Sandtiger Shark	0.59%	Angel Shark	0.00%
Night Shark	0.14%	Big-Eye Sixgill Shark	0.00%
Finetooth Shark	0.13%	OTHER	0.06%

Source: Morgan & Burgess, 2005[1]

2005

Life Histories and Vulnerability to Exploitation of
Elasmobranchs: Inferences from Elasticity, Perturbation and
Phylogenetic Analyses
Journal of Northwest Atlantic Fishery Science, 35: 27-45

FRISK, M.G. & MILLER, T.J. & DULVY, N.K.

We used life history traits to categorize vulnerability of
elasmobranchs to exploitation. However, the utility of this approach
required that the links between life histories and population dynamics be
explored. We constructed standardized three-stage matrix models for 55
species of sharks and rays. Using these models we (1) conducted elasticity
analyses to determine how the vital rates of mortality (M) and fertility (f)
influence elasmobranch population growth rate r, (2) determined the
response of elasticity to changes in the levels of exploitation, (3) estimated
sensitivity of elasticity to perturbation in vital rates, and (4) examined the
taxonomic distribution of model inputs and species vital rates, such as size
at maturity (L_{mat}), and total length (L_{max}). We found positive relationships
between the elasticity of λ (population growth rate) to changes in juvenile
and adult stages to longevity and age of maturity; however, the age of
maturity and the elasticity of λ to changes in the adult stage relationship
appeared to be invariant. There was a negative relationship between both
longevity and age of maturity and the elasticity of λ to changes in inter-
stage transitions of the models. Under varying fishing levels, estimates of
elasticity were robust to changes in survival. Elasticity and perturbation
analyses suggested that compensatory responses to exploitation in
elasmobranchs were less likely to be expressed as changes in fertility than
as changes in juvenile and adult mortality and stage durations.

Elasticity analyses are important in determining which aspects of a species' life history (fertility, juvenile survival, etc.) are significant in the face of perturbation: temporary disturbance inducing pronounced ecological change. The perturbation of fisheries pressure is of serious concern for sharks and rays. According to Frisk et al.'s model, an elasmobranch's survival into adulthood (G2) is the most sensitive life history aspect in the face of fisheries perturbation: as such, conservation for most species—including *Carcharhinus leucas*—should be focused on, "efforts to increase juvenile and adult survival that would have the greatest impact on population protection." In order to create a life history model (and thereby generate an elasticity analysis) for *Carcharhinus leucas*, Frisk et al. utilized the following species parameters:

Minimum Fecundity	F_{min}	6	Longevity	T_{max}	27
Maximum Fecundity	F_{max}	12	Age of Maturity	T_{mat}	15
Average Fecundity[1]	F_{ave}	2.25	Natural Mortality	M	0.17
Period Between Egg Pro.	F_{period}	2 yrs	Egg/Neonate Survival	M_1	0.00

1: F_{ave} = mid point between F_{min} and F_{max} adjusted for yearly reproductive cycle assuming a 50:50 sex ratio

On the continuum of elasmobranch life histories, *Carcharhinus leucas* is well seated within the group of large livebearers that exhibit matrotrophy—the strategy requiring, "additional maternal input through placentation, uterine milk or oophagy". Frisk et al.'s MDS ordination (a visualization of 'similarity' between elasmobranch life histories and elasticities) places *Carcharhinus leucas* close to *Negaprion brevirostris*: the Lemon Shark. This is rather interesting, given that the two species share a broad salinity tolerance and similar physique: according to the authors, "life history and demographic patterns are phylogenetically constrained." Do these two have a special relationship?

2005

Mitochondrial Gene Sequences Useful for Species
Identification of Western North Atlantic Ocean Sharks
Fishery Bulletin, 103 (3): 516-523

GREIG, T.W. & MOORE, M.K. & WOODLEY, C.M. &

QUATTRO, J.M.

Molecular-based approaches for shark species identification have
been driven largely by issues specific to the fishery. In an effort to establish
a more comprehensive identification data set, we investigated DNA
sequence variation of a 1.4-kb region from the mitochondrial genome
covering partial sequences from the 12S rDNA, 16S rDNA, and the
complete valine tRNA from 35 shark species from the Atlantic fishery.
Generally, within-species variability was low in relation to interspecific
divergence because species haloptypes formed monophyletic groups.
Phylogenetic analyses resolved ordinal relationships among
Carcharhiniformes and Lamniformes, and revealed support for the families
Sphyrnidae and Triakidae (within Carcharhiniformes) and Lamnidae and
Alopidae (within Lamniformes). The combination of limited intraspecific
variability and sufficient between-species divergence indicates that this
locus is suitable for species identification.

Greig et al.'s research on mtDNA proposes a phylogeny that aligns *Carcharhinus leucas* most closely with *Carcharhinus brevipinna*: the Spinner Shark. These two species are hypothesized to be monophyletic with *Carcharhinus isodon* (Finetooth Shark), *Carcharhinus acronotus* (Blacknose Shark), and *Carcharhinus porosus* (Smalltail Shark). Phenotypically, all five of these species share a common and unique identifier: the lack of an interdorsal ridge. This physical concordance with Greig et al.'s genetic findings may be testament to a true phylogeny between these five 'unridged' *Carcharhinus*.

However, we must remember our prior speculation concerning Frisk et al.'s life history study (where the Lemon Shark was aligned most closely to *Carcharhinus leucas*). None of the unridged *Carcharhinus* share as close a life history (and elasticity) to the Bull Shark as *Negaprion brevirostis*. Furthermore, *C. acronotus*, *C. brevipinna*, *C. falciformis*, *C. porosus* are all comparatively quite different from *Carcharhinus leucas* in both physique and habitat use: while the latter is large, heavy and euryhaline, the former are small, slight, and mostly restricted to full seawater. This is in fascinating contrast to the Lemon Shark, which, like the Bull, tolerates lower salinities, and bears a curious resemblance with its blunt snout, broad fins, and similarly heavy body.

This odd mismatch of potential relationships may uphold Greg et al.'s affirmation that, "the family Carcharhinidae was poorly supported as monophyletic". Intriguingly, the authors' neighbor-joining algorithms suggest that *Negaprion* is actually a "nested genus" within *Carcharhinus*; it must be cautioned, however, that this relationship has, "received little support from bootstrapping". Nevertheless, the prospect might hint at a complex evolutionary puzzle involving all six species: is there a direct link between *C. leucas*, the four unridged *Carcharhinus* and *Negaprion*?

2005

An Extremely Tolerant Shark: Osmoregulation in the Bull Shark, Carcharhinus leucas Abstract

American Elasmobranch Society 21st Annual Meeting, Tampa, Florida

MEISCHKE, L. & CRAMB, G.

The bull shark, Carcharhinus leucas, is a true champion of salinity tolerance. It can survive in water ranging from 0 to 53 ppt by precisely controlling the composition of its body fluids. This is achieved by the expression, function and coordination of several ion transporters and channels including the NA, K-ATPase. Bull shark Na, K-ATPase alpha and beta subunit isoforms have been amplified, cloned and sequenced. Northern blotting has been used to quantify messenger RNA expression in the rectal gland, gill, kidney and intestine of both FW- and SW-acclimated sharks. Using antibodies raised to the known sequences of both subunits, protein expression of Na, K-ATPase alpha and beta subunits have been compared in the gill, kidney and intestine. In addition, immunohistochemistry has been used to show the distribution of Na, K-ATPase within the osmoregulatory tissues.

It's funny to consider that the intricacies of osmoregulation—from the most basal molecular level of transporters, channels and isoforms, to the grander synchronization of gills, kidneys and intestines—are completely unapparent to *Carcharhinus leucas*. These are miracles in chemical conversion, and yet, they occur seamlessly; a legion of proteins can burn furiously to combat the deadly changes in salinity, and yet, remain quiet—tactful, in fact—so as not to bother their colossal master.

Osmoregulation is complex, vital, but most importantly, natural. It is a process akin to breathing—maddeningly complex in structure, and yet, marvelously simple in execution. Neither the Bull Shark nor we are quite conscious of our power to manipulate—to control the obstacles of our respective mediums. Every sweep of water into the gills—every breath of air into the lungs—is a proposal: are we capable of handling this influx? When we are delivered the essential gifts of oxygen, water, and salt—the electrifying cores of our being—we are encumbered with the challenge to receive: these marvelous materials must be processed quickly, efficiently and delicately, lest we face our own destruction. Every piece in our cellular complex must be nearly perfect, in order to safely synchronize; if one mechanism is to falter, it may jeopardize the entire system. Cells may explode. Salts may poison. Air may escape, and our minds would collapse. The gravity is real—the severity, enormous—and yet...

We are nonchalant. Breathing is easy. Salinity means nothing. Second after second, I perfectly prepare my breath. Second after second, the Bull Shark perfectly prepares its freshwater ascent: no mind, no bother—a seamless transition, from the ocean's merciless bereavement of water to the river's remorseless deprivation of salt. *Carcharhinus leucas* shares the disposition of the consummately adept: it pays no mind.

2005

Incorporating Variability in Size at Birth into Growth Models for Elasmobranchs: Does it Make a Difference?
AES 21st Annual Meeting, Tampa, Florida
NEER, J.A.

In order for age models to be considered accurate regarding the growth dynamics of a species, multiple models or various formulations of the same model may be required to determine which most accurately describes the growth of that species. Historically, the von Bertalanffy growth model (VBGM) with a to parameter has been the model applied most to elasmobranchs. More recently, some studies have begun to apply modified versions of the von Bertalanffy growth model. An alternate model introduced by Fabens (1965) reparameterizes the VBGM by removing the to parameter and forcing the model through the Y-intercept (e.g., hypothesized size-at-birth). While this model may be more applicable when there is an inadequate sample of very small individuals, the model still relies on one estimate of size-at-birth when, in reality, size-at-birth varies. To address the issue of variability in size-at-birth, a Monte Carlo simulation was incorporated into the size-at-birth intercept. The details of the methodology will be presented using the bull shark, *Carcharhinus leucas*, as a case study. Results of the analysis will be discussed for the bull shark, as well as several other species such as the finetooth shark, *C. isodon*, blacktip shark, *C. limbatus*, Atlantic sharpnose shark, *Rhizoprionodon terraenovae*, and the bonnethead shark, *S. tiburo*. This study provides the first attempt to incorporate variability at size-at-birth and provide measures of variability around the individual parameter estimates for elasmobranchs.

Neer's oral presentation for the American Elasmobranch Society's 21st Annual Meeting comes in tandem with her 2005 publication, *Age and Growth of Carcharhinus leucas in the Northern Gulf of Mexico: Incorporating Variability in Size at Birth* (Journal of Fish Biology, Volume 67, Issue 2, Pages 370-383). I imagine that the models, simulations and data of the AES presentation greatly overlap with those of this latter publication. Unfortunately, Neer's research is not publically available—that is, not without charge: access to this kind of data is only granted via purchase.

Here we arrive at an important crux of the scientific community: the price of knowledge. It's not unusual for scientific papers to require either a reading fee or full subscription to a journal for access; most of the larger publications charge about $30.00[3] apiece for a single paper, and anywhere between $50.00 and $20,000.00[3, 4] a year for access to a single journal. As for researchers, the cost to submit their own bodies of work can range anywhere from $300.00 to $5,000.00[2] (depending on the size, quality, and reputation of the scientific publisher to which they submit).

This is a controversial system. One on the one hand, publishers do require overhead, and need to generate an appropriate sum to cover the costs of publication (printing, editing, distribution, etc.). On the other hand, the profit margins can range from 15% to 50% for empirical journals[2]; this is a point of issue for many scientists, academics and institutions, who in some cases cancel their subscriptions entirely, and refocus their energies on leaner, more economical solutions to publication and secondary research. This latter dynamic creates an organic bias towards 'free knowledge' and away from 'costly knowledge', but this oddly undermines the spirit of science itself: to maximize objectivity, one must maximize accessibility. Are costs corroding the trade of thought?

2005

Characterizing Bull Shark (Carcharhinus leucas) Assemblages Near the Sabine Pass Inlet
Gulf of Mexico Science, 23 (2): 172-178
SHIPLEY, J.B.

The developmental stages of bull shark (*Carcharhinus leucas*) life history and the impact of selected environmental variables on the utilization of a Gulf of Mexico habitat by this species were characterized during late spring through summer 1992-1999. Entanglement nets 91.4 m in length of varying depth (2.40-4.88 m) and mesh sizes (12.7-25.4 cm) were deployed adjacent to jetty and beachfront sites near Sabine Pass. Bull sharks (N = 720) were incidentally captured as part of a study to monitor the population of Ridley sea turtles. The bull shark bycatch portion of the parent study data was expanded in 1997-1999 to record sex and in 1999 to include total length (TL) of individual bull sharks. Bull shark life history stages were estimated for the 1999 study from length and sex. Bull shark TL data when evaluated using size ranges of the *Final Fisheries Management Plan for Atlantic Tunas, Swordfish, and Sharks* indicate that no adult sharks were captured. Total length frequency compared to generally accepted length at age data supports that 94% of the Sabine Pass captures would be at most 6 yr old. A strongly correlated power model ($r^2 = 0.91$) extended the length-weight relationship data for immature life history stages of bull sharks. Bull shark catch (1992-1999) was positively correlated with water temperature (20.0-40.0°C), salinity (12.3-34.8 parts per thousand), and water clarity (0.0-1.6 m) and inversely correlated with dissolved oxygen (4.4-9.1 mg/liter).

Sabine Pass is an estuarine conduit of Sabine Lake, located on the border between Texas and Louisiana. The waters are warm, murky and shallow—a suitable setting for such a high yield of 720 *Carcharhinus leucas*. Indeed, Shipley's Sabine Pass dataset is one the most robust in Bull Shark literature—strong enough to support the author's conclusion that, "Sabine Pass can be considered an essential early life history bull shark habitat as defined by the Magnuson-Stevenson Act (NOAA, 1999)".

The catch of *Carcharhinus leucas* demonstrated moderately strong correlations with dissolved oxygen content ($r = 0.675$), water temperature ($r = -0.634$) and salinity ($r = -0.758$); however, the correlation with water clarity ($r = -0.846$) proved to be strongest. Shipley reports that turbidity, "probably enhanced bull shark capture by reducing their ability to see and avoid entanglement nets." In support of this affirmation, the author cites Wallace, 1972[5], who documented the ability of captive Bull Sharks to, "discriminate between nets of differing colors in tanks." It's fascinating to thus consider the power of visual obfuscation on this species (a shark that can normally distinguish between colors and textures). This knowledge might be quite useful to employ for future research surveys targeting *Carcharhinus leucas*; conversely, it also may be quite dangerous to share, lest anglers apply it to increase their own trophy catch of the mighty Bull.

Shipley's strongest contribution is arguably her length-weight model:

$$\text{Weight} = 1 \times 10^{-4}\text{Length}^{2.406}, \text{ where } r^2 = 0.9095$$

This model is the first of its kind to be generated for younger Bull Sharks. In addition, Shipley's documentation of skewed sex ratios marks an important deviation from the established 'nursery norm' of 1:1. Males consistently outnumbered females in the Sabine Pass study area, with a minimum ratio of 1.2:1 and a maximum ratio of 2:1.

2005

Elasmobranchs of Everglades National Park Abstract
AES 21ˢᵗ Annual Meeting, Tampa, Florida
WILEY, T.R. & SIMPFENDORFER, C.A. & TYMINSKI, J.P.

Few directed studies have been conducted on the elasmobranch fauna of Everglades National Park, so little is known of their importance in this area. Data on elasmobranchs captured during surveys for smalltooth sawfish in Everglades National Park were used to examine species occurrence, distribution and movement patterns. Surveys were conducted utilizing bottom set longlines, gillnets, seine nets and rod and reel from July 2000 to February 2005. A total of 1015 elasmobranchs of 12 species were identified within the Park (*Carcharhinus acronotus, C. isodon, C. leucas, C. limbatus, Dasyatis* spp., *Galeocerdo cuvier, Ginglymostoma cirratum, Negaprion brevirostris, Pristis pectinata, Rhizoprionodon terraenovae, Sphyrna mokarran* and *S. tiburo*). *Carcharhinus leucas* (n=302), *N. brevirostris* (n=239), *C. limbatus* (n=126) and *G. cirratum* (n=112) were encountered most frequently. Data from Mote Marine Laboratory's tag/recapture database showed that 20 *C. leucas*, three *C. limbatus*, eight *G. cirratum*, 16 *N. brevirostris* and two *S. tiburo* were tagged and/or recaptured within the park after periods at liberty of 1 to 1099 d. All 16 *N. brevirostris* were tagged and recaptured within the boundaries of the park (44 to 1061 d at liberty). *Carcharhinus leucas , C. limbatus* and *G. cirratum* were tagged or recaptured as far north as Tampa Bay on Florida's west coast. No recaptures were reported from Florida's east coast or south of the park boundary. The salinity, temperature, depth and dissolved oxygen data were analyzed to determine any environmental preferences.

The Everglades are by now an infamous system. Local development and exotic invasion have engulfed the "River of Grass" with violent and fiery change. Politicians, civilians, industries—all cogs in the urban machine—have, for over a century, stripped the forests of their wood and deprived the wetlands of their water. The human thirst for profit and need for water have denuded the original landscape, and by now, only 20% of the original Everglades remain within the protection of Everglades National Park. Even so, the system feels the bereavement of freshwater—redirected to growing urban sectors like Miami—and the indulgence of invasives: Walking Catfish (*Clarias batrachus*), Burmese Python (*Python molurus*) and now Nile Crocodile (*Crocodylus niloticus*—currently present, but not established) have destroyed the integrity of the original ecosystem, and threatens natives like the Florida Panther with local extinction.

From this disastrous complex emerge the nutrients and freshwaters of the Everglades. Slowly these seep into the salty mangroves of Florida Bay to create a rich (but perilously afflicted) estuary. It is from this habitat that twelve species of shark draw food, water, oxygen and protection. It is from this habitat that *Carcharhinus leucas* expects a living. Here, the Bull Shark is king—the dominant large marine predator. Florida Bay and the Everglades are not simple, run-of-the-mill habitats: they are strongholds— crown jewels of this species' American empire. These epicenters place *Carcharhinus leucas* at the top of abundance—a seemingly surreal position, given that the shark is typically listed as an uncommon vagrant within most American surveys.

The Great Question is one of security: will *Carcharhinus leucas* be *allowed* to keep the Everglades and Florida Bay? By nature, its right is divine—by man, in debate.

King for now, but on a corroding throne...

2006

Sharksucker-Shark Interaction in Two Carcharhinid Species
Marine Ecology, 27: 89-94
BRUNNSCHWEILER, J.M.

It is not known whether sharksuckers have positive or negative effects on their hosts, partly because this association is difficult to study in free-ranging fish. I observed the behavior of sharks with and without sharksuckers, to determine whether the hosts actively avoid sharksuckers. Wild blacktip sharks, *Carcharhinus limbatus*, took evasive actions when sharksuckers, *Echeneis naucrates*, attached to them, presumably to escape from skin irritation or hydrodynamical drag caused by the sharksuckers. Sharksuckers were most often attached to the belly or back of the shark, and sharks reacted most strongly to sharksuckers on their heads, sides, and dorsal fins. Observations of two captive bull sharks, *Carcharhinus leucas*, indicated that swimming speed increased when sharksuckers were attached. This paper supports the hypothesis that sharksucker attachment irritates sharks, and that the relationship between the two is best viewed as a subtle host—parasite interaction.

The remoras are a disarmingly fascinating group, famed for their bizarre (and slightly grotesque) modified dorsal sucker, which, in mythology, has been attributed to the disruption of many a nautical battle. According to Pliny the Elder, it was the remora that secured Mark Antony's defeat at the Battle of Actium in 31 BCE: too many had attached to the hulls of Antony's ships, and the General fell victim to their monstrous drag—a fatal blow to tactical maneuverability. Pliny goes on to imply that remoras also had a hand in the assassination of Emperor Caligula in the year 41: during his Imperial Majesty's return from Gaul, so many had attached to his vessel that he became delayed, and therefore, susceptible to the expedience of his assassinators.

From this rich tradition of impediment and countermine, some remoras have emerged into the Age of Science with a noble genus: *Echeneis*—from the Greek, "ship-holder". Amusingly, Pliny's weighty accusations against these adhesive abhorrences bear little historical or biological potency: it is nearly impossible for a remora to literally forestall a ship. However, the species *Echeneis naucrates*—the Sharksucker— possesses at least some powers of prohibition, as attested by this study.

For *Carcharhinus leucas*, Brunnschweiler reports that the Sharksucker is an irritant, forcing its host to increase swimming speed and physically protest with sharp turns, pitches and various wiggles (perhaps in an attempt to dislodge the persistent 'semi-parasite'). Certain areas of the shark's body are said to be more sensitive to the infamous disk of *Echeneis* than others: the head, gills, pectoral and first dorsal fins are, according to the author, generally touchier than the back, belly and tail. Attachment to the lateral line may too be a source of discomfort, but this is uncertain. More research is needed to fully understand the Tortures of *Echeneis*; are they truly the bane of a Bull Shark?

2006

A Contribution to Marine Life Conservation Efforts in the
South Pacific: The Shark Reef Marine Reserve, Fiji
Cybium, 30 (4) (Supplement): 133-139
BRUNNSCHWEILER, J.M. & EARLE, J.L.

ABSTRACT. – The first estimate of the fish biodiversity of Shark Reef
Marine Reserve (SRMR), Fiji, is provided with special emphasis on
elasmobranchs. In 2004, nine elasmobranch species were regularly
observed at the site. The most common were the bull shark *Carcharhinus
leucas* and the grey reef shark *Carcharhinus amblyrhynchos*. During the
fish count made in fall 2004 a total of 267 species of fishes were recorded
at SRMR, including members of 37 families, the most diverse being that of
the Labridae. SRMR could serve as a model to implement the local marine
conservation efforts; this can only be achieved by the creation of an
integrated management system taking into consideration all human
activities and their impacts.

Key words. – Sharks – Elasmobranchii – ISEW – Fiji – Fish Biodiversity –
Marine Protected Areas.

Carcharhinus leucas is the main attraction of the Shark Reef Marine Reserve: a marine protected area located between the southern coast of Viti Levu—Fiji's largest island—and the Beqa Lagoon. From January to May, Bull Sharks are abundant enough to create a world-class diving experience featuring *Carcharhinus leucas* specifically (making the SRMR a global hotspot for Bulls). As an anchor for the local dive economy, the SRMR serves as an exemplary conservation area, not only for its protection of valuable marine habitat (from local fishing pressure), but also for its accomplishment of, "ecological, economic, and social objectives such as recreation, education, and research".

With their survey, Brunnschweiler & Earle aim to provide a baseline of abundance, so as to facilitate any continued assessment of the SRMR's impact on local biodiversity. From survey dives, the authors recorded 267 fish species—an impressive number, considering that only seven dives were conducted during the survey period. Alongside *Carcharhinus leucas*, seven other sharks were reported within the Reserve: *Negaprion acutidens* (Sharptooth Lemon Shark); *Galeocerdo cuvier* (Tiger Shark); *Carcharhinus albimarginatus* (Silvertip Shark); *Nebrius ferrugineus* (Tawny Nurse Shark); *Carcharhinus amblyrhynchos* (Grey Reef Shark); *Carcharhinus melanopterus* (Blacktip Reef Shark); and *Triaenodon obesus* (Whitetip Reef Shark).

Though Brunnschweiler & Earle do not address it, there could be a fascinating seasonal dynamic taking place between *Carcharhinus leucas* and *Carcharhinus amblyrhynchos*: as Bull Sharks decrease in number from June to December, Grey Reef Sharks increase in number. Could this be a sort of local 'predator-release' wherein the Bull Shark—the theoretically dominant predator— yields its apex position during its seasonal (and purportedly breeding) migrations out of the SRMR?

2006

Exploring Intraspecific Life History Patterns in Sharks
Fishery Bulletin, 104 (2): 311-320
COPE, J.M.

Marine ecosystems compose the major source (85%) of world fisheries production. Although only a few fish species tend to dominate fishery catches, a large diversity of fishes representing varied taxonomic levels, ecological guilds, and life histories is commonly taken. Recently, 66% of global marine resources were determined to be either fully, heavily, or over-exploited. Considering the current state of many fisheries, the large diversity of species taken globally, and the general lack of resources to adequately assess many stocks, it has become important to develop shortcuts that may provide methods fisheries scientists can use to determine which stocks are in danger of overexploitation and which recovery plans are appropriate when biological data are limited.

Applications of life history theory have proven a potentially useful means to accomplish such tasks. Life history traits such as maximum size and age, maturity, mortality, and growth are correlated among teleost fishes and relationships among such traits can be used to infer some general life history patterns. These general patterns reveal that teleost fishes wth higher maximum ages tend to be larger, mature later, grow more slowly, and have lower natural mortality rates (K-selected species), whereas teleost fishes with lower maximum ages tend to show the opposite relationships (r-selected species).

'World Ocean' is an appropriate designation for the vast, literally endless expanse of saltwater covering 71% of the planet's surface; simultaneously, it is also a terrible misnomer. With 'World Ocean', one is lured into a false sense of uniformity—an image of an unchanging sea. However, in truth, there is monstrous variation within and between oceanic basins, as each possesses a unique combination of geography, climate, and hydrodynamics: Atlantic, Pacific, Indian, Caribbean—these are not merely names on a map, but entirely original, peerless identities, each with a distinct ecological impression.

In light of this brilliant truth, Cope asks a powerful question concerning cosmopolitan shark species: how does intraspecific (that is, "within a single species") life history vary between different oceanic basins? For *Carcharhinus leucas*—one of Cope's seventeen globally distributed species—there is a difference in both size and fecundity between populations of the Gulf of Mexico and populations of the Indian Ocean:

	Max. Size	Mature Size (Female)	Mature Size (Male)	Mean Fecundity
Gulf of Mexico	285.0 cm	225.0 cm	215.0 cm	8
Indian Ocean	300.0 cm	239.5 cm	239.5 cm	9

For Bull Sharks, it appears that the Indian Ocean is a more fruitful basin than the Gulf of Mexico. Only one other species shares this particular difference in life history patterns: *Carcharhinus limbatus* (the Blacktip Shark). Conversely, many species conform to an amazing global trend: that, "across taxonomic designations, populations in the northern latitudes tended to be larger, to mature later in life, to have longer life spans, and to have greater fecundity compared to conspecifics in the central and southern latitudes." Populations of the Northern Pacific, in particular, are superlative among conspecifics.

2006

Further Notes on the Capture of a Carcharhinus leucas in a Northeastern Atlantic Oceanic Insular Shelf, the Azores Archipelago, Portugal

Cybium, 30 (4) (Supplement): 31-33

GADIG, O.B.F. & JULIANO, M.F. & BARREIROS, J.P.

In March 1993, a specimen of *Carcharhinus leucas* was captured by fishermen on the south coast of Terceira Island, the Azores Archipelago. Its head was recovered and its jaws were preserved. This is the first capture of this species on an oceanic insular shelf in the Atlantic. The distribution of *C. leucas* in this ocean is commented.

The Azores archipelago, northeastern Atlantic, is a group of nine volcanic islands situated on the Mid-Atlantic ridge. There are few shark studies in the area, most of them are included in other works related to a commented list of the entire fish fauna from the Archipleago. In the present study, the occurrence of the bull shark, *Carcharhinus leucas* (Valenciennes, 1839), is reported from the Azorean waters, based on a single specimen caught in March 1993 in São Mateus Bay, Terceira Island (38°39'N; 27°13'W) (Fig. 1). Although this specimen has been referred in a previous checklist of Azorean marine fishes (Santos *et al.*, 1997) this note provides available information on this regard with comments on geographical distrubtion of this species.

The shark was sighted swimming in shallow waters and was caught by a fisherman using a hand harpoon near the surface. The shark was eviscerated and its meat commercialized while the carcass was discharged in the bay, where its head (Fig. 2) was recovered by JPB 12 hours later.

The provisional number for this particular specimen of *C. leucas* is: CARLEU 1/94/05. Though only the head and jaws were conserved, the authors could still estimated 'CARLEU's total length as 250-270 cm (based on the systematic precedent that, in *Carcharhinus leucas*, the mouth width corresponds to roughly 9.5 to 10% of the total length). CARLEU represents the Eastern North Atlantic's northernmost record of *C. leucas*, and furthermore, the world's, "highest latitudinal (South or North) record of this species in an oceanic island". Previous records of this species in Europe are under much speculation, as many have been misappropriations involving the Dusky Shark (*C. obscurus*) or the Sandbar Shark (*C. plumbeus*).

According to Gadig et al., Bull Sharks are not known to occupy the oceanic islands and insular shelves of the Atlantic. CARLEU's occurrence in the Azores is revolutionary, and brings into question the Gulf Stream affect on this species' distribution. Gadig et al. suggest that the Azores Current (an offshoot of the Gulf Stream) may seasonally transport tropical fauna from the Western North Atlantic (particularly the Caribbean Sea) to the Azores during summer in the northern hemisphere. This hypothesis is supported by the fact that Bull Sharks typically occur in, "coastal areas that border the external circulation path of the subtropical gyre in the west north Atlantic."

In additional to these possible seasonal effects on distribution, the authors affirm that CARLEU was captured in 1993: an anomalous year for both the Gulf Stream and the Azores Current. From year to year, oceanic surface currents can vary in both position and energy, and from 1993 to 1995, the Gulf Stream shifted northward, while, "the eddy kinetic energy associated to the Azores Current reached its peak". Given such compelling forces, it is quite possible that the Bull Shark also occurs in the Canaries and Madeira.

2006

Analisi Sistematica, Paleoecologica e Paleobiogeografica
della Selaciofauna Plio-Pleistocenica del Mediterraneo
UNIVERSITÀ DI PISA, Doktorarbeit
MARSILI, S.

The systematic, paleoecologic and paleobiogeographic analysis of the Plio-Pleistocene Mediterranean elasmobranch fauna herein discussed provides new relevant data concerning the diversity and disparity of this important marine vertebrate group, in relation to the main climatic, oceanographic and geological evolutionary events. Moreover, This study is part of a wider project in order to understand the main Neogene historical and evolutionary processes involved in the establishment of the present Mediterranean fish fauna.

The database used, comprehensive of all the Plio-Pleistocene shark teeth fauna of the Mediterranean area, was compiled from a generic and species revision of historical elasmobranch teeth collections, from a study of the new findings collected the middle Pliocene sections of Rio Merli, Rio dei Ronchi, Rio Cugno and Rio Co di Sasso, and from the lower-middle Pleistocene sections of Fiumefreddo, Grammichele, Archi and Vallone Catrica, as well as from a critical analysis of as many publications as possible figuring and describing Plio-Pleistocene Mediterranean elasmobranch faunas. The qualitative and quantitative analysis of this database provided a general paleoecological and paleobiogeographical pattern of the Mediterranean elasmobranch fauna through the Plio-Pleistocene.

Marsili's dissertation includes a comprehensive stratigraphic record of fossil *Carcharhinus leucas* from across the globe. Here is a summation of these deposits:

EPOCH	LOCATION	AUTHOR	YEAR
Pleistocene	Florida	Scudder et al.	1995
Pleistocene	Japan	Uyeno & Matsushima	1974
Pleistocene	Celebes, Indonesia	Hooijer	1954
Pliocene	Lee Creek Mine, North Carolina	Purdy et al.	2001
Pliocene	Luanda, Angola	Antunes	1963
Pliocene	Italy	Bellocchio et al.	1991
Miocene	Maryland	Agassiz	1833-43
Miocene	Maryland	Leriche	1942
Miocene	North Carolina	Purdy et al.	2001
Miocene	Ecuador	Longbottom	1979
Miocene	Portugal	Antunes et al.	1999
Miocene	Portugal	Antunes & Balbino	2004
Miocene	Malta	Menesini	1974

In Italy, fossil *Carcharhinus leucas* from the lower-middle Pliocene have been mistakenly identified as *Carcharhinus egertoni*: a 'wastebasket taxon'. Marsili corrects this error by reassigning several *C. egertoni* teeth as either: *C. brachyurus*; *C. falciformis*; *C. leucas*, *C. longimanus*, *C. obscurus*, *C. perezi* or *C. plumbeus*. The Mediterranean once boasted a high diversity of *Carcharhinus* in the Pliocene, but this has since decreased with the disappearance of tropical and subtropical taxa like *C. leucas*.

2006

Economically Important Sharks and Rays of Indonesia
ACIAR Monograph Series, 124: 1-329
WHITE, W.T. & LAST, P.R. & STEVENS, J.D. & YEARSLEY, G.K. & FAHMI, D. & DHARMADI

The Republic of Indonesia is the most expansive archipelago in the world, consisting of more than 17,000 islands, most of which are uninhabited. Indonesia straddles the equator and stretches for almost 5,000 km from Sabang in northern Sumatra (5°38'N, 94°44'E) to near Merauke on the Papua-New Guinea border (141°37'E), and for 1,770 km south to the island of Roti off West Timor (13°33'S). It has a tropical climate and consists of five main islands, namely Sumatra, Java, Kalimantan, Sulawesi and Papua. The physical seascape of Indonesia is unique, consisting of a complex series of shelves, volcanic mountain chains and deep-sea trenches. The geological history is also complex, which has resulted in Indonesia having the highest rate of marine endemism in the world. It contains the most diverse seagrass meadows in the world (with more than 12 species), the greatest expanses of mangroves, and extensive coral reef communities (covering an area of more than 75,000 km^2).

The extremely high diversity of marine habitats has yielded one of the richest finfish and elasmobranch faunas in the world. The declaration of the Indonesian Exclusive Economic Zone (EEZ) in 1983, which includes FAO Fishing Areas 57 and 71, ensures that Indonesia maintains control of more than 5.4 million km^2 of marine waters. Subsequently, Indonesia's fisheries are large.

Indonesia has the world's largest (and possibly most biodiverse) chondrichthyan fishery, and yet, "almost no published biological data or size compositions of species landed." In response to this egregious lack of information, the Australian Centre for International Agricultural Research (ACIAR) spearheaded a collaborative effort in 2001 to, "survey the chondrichthyan landings at various fish markets in eastern Indonesia". A total of 850 specimens representing 78 sharks, 56 rays and 3 chimaeras were catalogued by the time of this unprecedented field guide's publication in 2006.

The Bull Shark—known locally as Hiu Buas or Cucut Bekeman— is one of 17 *Carcharhinus* and 27 Carcharhinids reported by the ACIAR to inhabit Indonesian waters. Here, the shark is caught occasionally by longline, tangle net and gillnet fisheries, primarily for the purpose of acquiring fins, which are, "high value in adults". White et al. provide one of the most expansive descriptions of the Bull Shark's diet, stating that it is composed, "primarily of bony fishes and elasmobranchs, but also turtles, crocodiles, birds, dolphins, terrestrial mammals and crustaceans." The inclusion of crocodiles is rather interesting, as most studies omit these reptiles as a dietary option for *Carcharhinus leucas*. Could this hint at a heightened predatory-prey dynamic between the two hunters—a dynamic unique to Indonesian waters? After all, few other places boast a legend quite like that of Surabaya—a legend where the Bull Shark and crocodile famously engage in violent combat over food and territory; this immortal tale may very well be, in tandem with other aims, a cultural reflection of natural observation.

White et al. make a curious choice in referring to the species as, "an extremely dangerous shark to humans." The superlative 'extremely' seems rather unwarranted, given this species docility in ecotourism efforts.

2006

Widespread Utility of Highly Informative AFLP Molecular Markers across Divergent Shark Species
Journal of Heredity, 97 (6): 607-611
ZENGER, K.R. & STOW, A.J. & PEDDEMORS, V. & BRISCOE, D.A. & HARCOURT, R.G.

Population numbers of many shark species are declining rapidly around the world. Despite the commercial and conservation significance, little is known on even the most fundamental aspects of their population biology. Data collection that relies on direct observation can be logistically challenging with sharks. Consequentally, molecular methods are becoming increasingly important to obtain knowledge that is critical for conservation and management. Here we describe an amplified fragment length polymorphism method that can be applied universally to sharks to identify highly informative genome-wide polymorphisms from 12 primer pairs. We demonstrate the value of our method on 15 divergent shark species within the superorder Galeomorphii, including endangered species which are notorious for low levels of genetic diversity. Both the endangered sand tiger shark (*Carcharodon taurus*, $N = 18$) and the great white shark (*Carcharodon carcharias*, $N = 7$) displayed relatively high levels of allelic diversity. A total of 59 polymorphic loci ($H_e = 0.373$) and 78 polymorphic loci ($H_e = 0.316$) were resolved in *C. taurus* and *C. carcharias*, respectively. Results from other sharks (e.g., *Orectolobus ornatus*, *Orectolobus* sp., and *Galeocerdo cuvier*) produced remarkably high numbers of polymorphic loci (106, 94, and 86, respectively) from a limited sample size of only 2.

224

Zenger et al. assert that their method—an amplified fragment length polymorphism (AFLP) genetic marker system—can be an informative approach to, "individual DNA fingerprinting, genetic diversity estimates, population structure analysis, identification of hybridization, and identifying adaptive traits and species phylogenies." Prior studies have utilized AFLP as an effective tool for, "detecting subtle but significant genetic subdivisions." This keen sense of discernment is particularly important for the assessment of many shark species and their genetic structuring, as sharks typically demonstrate, "low variability at allozyme loci and in mitochondrial DNA." Furthermore, AFLP has been proven to be a rather economical method of genetic marking as, "thousands of genome-wide bi-allelic dominant loci can be evaluated in a relatively short period of time and at a low cost."

Not much is said for *Carcharhinus leucas* specifically in this study. The family Carcharhinidae—here termed as the 'Whaler Sharks'—demonstrated a much higher proportion of shared AFLP bands within itself than with the other taxa (as to be expected). By extension, the highest proportions of shared bands were, "between the most closely related species in each group". Unfortunately, the authors do not elaborate as to which species *Carcharhinus leucas* most genetically resembles: *C. limbatus* (Blacktip Shark), *C. brachyurus* (Bronze Whaler), and *C. amboinensis* (Java Shark) were the possible candidates, given that they were the three other *Carcharhinus* analyzed. Based on an (uneducated) examination of the AFLP profile (using the primer combination E-AGT and M-CAAC) generated for all 15 divergent shark taxa, I suspect that *Carcharhinus amboinensis* bears the closest resemblance to *Carcharhinus leucas*. Such an estimate would also fit quite nicely with the species' phenotypic resemblance.

2007

First Report on the Philopatric Migration of Bull Shark,
Carcharhinus leucas in the Pulicut Lagoon
Marine Fisheries Information Service, Technical and
Extension Series, 191: 30
BATCHA, H. & REDDY, P.S.R.

Four pregnant bull sharks *Carcharhinus leucas* were captured by a modified gillnet, made of cotton twine from Pulicat lagoon during 2005-06. The bull sharks exhibit unique philopatric behavior in the lagoon for parturition.

The four specimens caught were all pregnant bull sharks, measuring 330, 300, 320 and 350 cm total length and weighed 325,200,320 and 335kg respectively. The bull shark is viviparous with yolk-sac placenta. Total length of sharklings varied from 620 to 840 mm with a weight of 3.5 to 4.1kg. The value of the sharks ranged from Rs. 20000-22000.

There are literally no other details given for this account. We are thus left to contemplate the Bull Shark' philopatry—the tendency to stay in, or habitually return to, a particular area—in Pulicat Lagoon. Is Pulicat a case of breeding-site philopatry, in which the Lagoon serves as a habitual nursery ground for *Carcharhinus leucas*? Or could this be an instance of natal philopatry, wherein each pregnant female (almost miraculously) returns to the site of her own birth? To lack such crucial information and inquiry is rather infuriating—even more so when were are left to be satisfied with the arousing claim that these Bulls, "exhibit unique philopatric behavior in the lagoon for parturition."

There is, however, much to consider when it simply comes to the idea of philopatry in itself. In any capacity, philopatry is an intimate relationship: the organism is not merely interacting with its environment, but *committing* to it. In a philopatric relationship, the environment must be stable, dependable and benevolent; it must never deviate from this position of trust, lest the organism's fidelity be compromised. In Pulicat, the Bull Shark makes a tremendous investment: the commitment to pup—to devote precious time and energy for propagation—and to not explore the prospects of other waters. This fidelity is quite literally exclusive, and the Bull Shark takes risk in losing its other options (regardless of their quality): if *Carcharhinus leucas* is truly philopatric with Pulicat, then Pulicat is partner with this species' existence.

In the Romantic sense, philopatry appeals to the ideals of 'home'. The perquisites of fidelity, stability and benevolence honor a perfected domestic essence. Bull Shark mothers, though not caregivers in the human sense, nevertheless channel these values in their philopatric behavior towards places like Pulicat. Their habitual haunting betrays a profound devotion: their commitment towards, what is essentially, the safest cradle.

2007

Origin and Diversification of the Neotropical Ichthyofauna: A Review
Cybium, 31 (2): 139-153
BRITO, P.M. & MEUNIER, F.J. & LEAL, M.E.C.

The neotropical ichthyofauna has approximately 8,000 living species, representing 25% of the "fish" diversity in the world, including marine and freshwater forms. Most of this diversity, at least for the freshwater species, is distributed in the vast South American continent. The history of this biodiversity started after the complete separation between South America and Africa, at the end of the Lower Cretaceous, approximately 100 million years ago. This spectacular evolutionary radiation resulted from a series of events of dispersion and vicariance which has occurred between early Upper Cretaceous and Late Miocene. The displacement of the South-American shield to the west and its collision with the Nazca plate led to important geographical changes such as the rise of the Andes and the inversion of the direction of the current of the principal hydrographic basins, like the Amazon, which originally was running to the west. The sea level oscillations related to many glacial periods in the Quarternary, were also important factors for the richness of continental fish fauna. Aside from the "typical" marine species that also can be found in freshwaters (e.g., *Carcharhinus leucas*, *Pristis pristis*, *P. pectinata*, *Megalops atlanticus*, *Mugil* spp.), or those that live in freshwater only during the reproduction and larval life (e.g. the lamprey, *Geotria australis*), the extant neotropical ichtyofauna can be divided in three main categories of taxa:

The three categories of taxa to which Brito et al. refer are: the "Gondwana taxa" (freshwater taxa that survived the ancient continental split of Gondwana into South America and Africa, and thereby populated both contemporary continents); the "marine taxa" (freshwater taxa that originated from a marine ancestor); and the "adaptive radiation taxa" (freshwater taxa that emerged from Neotropical sister groups of adjacent freshwater systems). Unfortunately, *Carcharhinus leucas* does not fall into any of these three groups, and is therefore not discussed in detail as a Neotropical freshwater feature.

Brito et al. instead paint a grander picture of the biodiversity of Neotropical freshwater ichthyofauna by invoking the massive geological and climatic events responsible for creating such a rich assemblage of life. During the Cretaceous Period, the supercontinent Gondwana split to form the modern landmasses of South America and Africa. This process was the first major dictation of Neotropical character; its magnitude can be affirmed by the substantial difference in biodiversity between African and South American freshwater ichtyhofauna (the fauna of South America are, in terms of biodiversity, far richer than that of Africa). By the Miocene Epoch, the Andes expanded to their full measure, and locked the eastward paths of most Neotropical rivers into place. The alignment of this continental vertebra—in tandem with fluctuating sea levels and shifting plates—carved an unimaginably complex skeleton for thousands of freshwater ecosystems, each partitioned with its own unique community and sense of biodiversity.

Though *Carcharhinus leucas* may not be intimate with such a powerful legacy, its freshwater access and predatory role no doubt shape the path of its progression. Like the continents, mountains and seas before it, *C. leucas* serves as one of many sculptors: with its ecological influence, it alters, even in the smallest way, the Neotropical destiny.

2007

Volumetric Analysis of Sensory Brain Areas Indicates
Ontogenetic Shifts in the Relative Importance of
Sensory Systems in Elasmobranchs
Raffles Bulletin of Zoology, Supplement 14: 7-15
LISNEY, T.J. & BENNETT, M.B. & COLLIN, S.P.

Studies on the brains of teleost fishes have shown that the relative size of sensory brain areas reflects sensory specialisations and the relative importance of a given sensory system. Moreover, the relative size of these brain areas can change in relation to ontogenetic shifts in habitat and feeding ecology. However, although elasmobranchs (sharks, skates and rays) also exhibit ontogenetic shifts in habitat and diet, little is known about how their sensory systems and brains may adapt to these changes. In this paper, we compare the relative volumes of four sensory brain areas; the olfactory bulbs, optic tectum, anterior lateral line lobes and posterior lateral line lobes (that receive input from the olfactory epithelium, eye, electroreceptors and lateral line, respectively) in juveniles and adults of seven species of elasmobranch (six species of shark and one species of ray). The relative volume of each brain area was expressed as proportion of the total sensory brain, the combined volume of the four brain areas. Significant differences were found in the relative proportions of the sensory brain areas between juveniles and adults. In all species, the optic tectum was relatively larger in juveniles, whereas the size of the olfactory bulbs was relatively greater in adults. This paper provides the first evidence for shifts in the size of sensory brain areas in elasmobranchs and suggests that vision is relatively more important than olfaction in juvenile elasmobranchs and vice versa in adults.

Lisney et al. sampled four juvenile *Carcharhinus leucas* and recorded an absolute mass of 241.73 ± 42.08 mm^3 for the olfactory bulbs (39.2 ± 1.7% of the total sensory brain) and 192.17 ± 23.58 mm^3 for the optic tectum (31.3 ± 1.0% of the total sensory brain). Only one adult Bull Shark was sampled, but its proportions were markedly different: from this individual, the authors recorded an absolute mass of 1,294.05 mm^3 for the olfactory bulbs (52.3% of the total sensory brain) and 518.27 mm^3 for the optic tectum (20.9% of the total sensory brain). As with other elasmobranchs, the Bull Shark's olfactory bulbs experience an ontogenetic increase in relative mass, while the optic tectum experiences an ontogenetic decrease; this may plausibly correlate to a shift in sensory dependence, as, "there is evidence that the relative size of these brain areas reflects the relative importance of their corresponding sensory modalities"

As such, it appears that with age, *Carcharhinus leucas* becomes more reliant on a keenness of smell than on sight—but why? One of the more compelling hypotheses relates to the species' use of coastal nurseries, where, "a greater range of wavelengths of light may be available in comparison to deeper waters occupied by adults". Juvenile Bull Sharks may, at first, capitalize on the broader light spectrum by relying more heavily on visual cues for both hunting and predator avoidance (the latter being an especially important concern for such a vulnerable size class). One they age, adult Bull Sharks become less concerned with predator avoidance and more so with the ability to track larger prey in deeper, more open waters; to find such far-ranging food, the sense of smell may need to heighten. Likewise, olfaction may be, "more important in sexually mature adults as pheromones are thought to play an important role in reproductive behaviour."

2007

Revision of the Teeth of the Genus Carcharhinus
(Elasmobranchii; Carcharhinidae) from the
Pliocene of Tuscany, Italy
Rivista Italiana di Paleontologia e Stratigrafia, 113 (1): 79-
96

MARSILI, S.

The great similar tooth morphologies that characterizd the sharks of the genus *Carcharhinus* have suggested that the Neogene Mediterranean Sea were inhabited by only one or two widespread carcharhinid taxa, *Carcharhinus egertoni* and *C. priscus*. Five new shark species included into *Carcharhinus*, *C.* aff. *brachyurus*, *C. falciformis*, *C. leucas*, *C. perezi*, and *C. plumbeus*, have been identified by the review of some shark teeth from Tuscany (Italy) and housed in the Italian Museums of the Universities of Firenze, Bologna, and Pisa. Moreover, the Mediterranean records of the two Recent *C. longimanus* and *C. obscurus* have been confirmed too. Paleobiogeographic and palaeoecologic analysis have been produced on the presence of some species absent or doubtful for the extant Mediterranean shark community.

Carcharhinus may have first appeared during the Middle Eocene (48 to 38 Ma), and has been recorded from this period in Egyptian, Moroccan and North American deposits (the world's earliest records for this genus). Though Oligocene (34 to 23 Ma) fossils also exist for this taxon in Belgium, it was not until the Miocene (23 to 5 Ma) that *Carcharhinus* became, "one of the most prominent elements of the neritic fauna." As such, numerous Miocene deposits are scattered throughout the Mediterranean (Italy, France, Libya and Algeria) and beyond.

The Miocene deposits of the Alvalade basin, Portugal and Malta are the first records of *Carcharhinus leucas* in the Mediterranean basin. Pliocene (5.3 to 2.6 Ma) fossils of this species have also been found in the Umbria region, Italy. Together, these distributions hint of a warmer Mediterranean past, and correspond with, "the benthic foraminifers, molluscs, pollen, geological and isotopic data, that suggest a warmer (by 5C), more humid (by 400-1000 mm/yr) and less seasonal clime than present". With the cool of the Pleistocene (2.6 to 0.1 Ma), *Carcharhinus leucas* appears to have shifted its distribution southward, as no Bull Sharks remain in the Mediterranean Sea today.

Mediterranean *Carcharhinus* fossils have been historically divided into two groups: the now familiar *C. egertoni* group and the *C. priscus* group. *Carcharhinus leucas* has been traditionally (and mistakenly) associated with the former thanks to the triangular shape, heavy serrations and straight (or slightly wavy) mesial cutting edge of their upper teeth. Though fossil *C. leucas* teeth can be easily distinguished from the narrower, more distally inclined cusps of the *C. priscus* group, the fellow members of *C. egertoni* (*C. falciformis* and *C. longimanus*) are harder to differentiate. This could have hinted at shared phylogeny, if not for the fact that these latter two are 'ridgeback' sharks.

2007

Tiburones y Rayas de Colombia (Pisces: Elasmobranchii)
Lista Revisada, Actualizada y Comentada
Boletín de Investigaciones Marinas y Costeras, 34: 128-169
MEJÍA-FALLA, P.A. & NAVIA, A.F. & MEJÍA-LADINO, L.M.
& ACERO, A.P. & RUBIO, E.A.

Check list of the species of sharks, rays and skates (Pisces: Elasmobranchii) registered in Colombian waters. A revision of the species of sharks, skates and rays registered for the marine and continental waters of Colombia was carried out; a total of 176 species (88 sharks and 88 batoids) were found. Of these, 63 species of sharks and 58 of skates and rays have been confirmed through photographic records and biological collections, and the remaining has only been supported by bibliographical references. This inventory includes 69 genera (36 sharks and 33 batoids) and 34 families (20 sharks and 14 batoids). Four new species and 20 new reports for Colombian waters have been published in the last 25 years. Six species are deposited in internationals museums but have not appeared in the scientific literature. The number of Colombian material cataloged and deposited in national and international biological collections and collections of images is also included, as well as the amount of existent Colombian bibliographical references on each species. Taxonomic and systematic comments of some groups which validity is discussed are made.

KEY WORDS: Elasmobranchs, Taxonomy, Caribbean, Pacific, Continental waters, Colombia

Thirty-eight separate records confirm *Carcharhinus leucas* as an estuarine-coastal inhabitant of the Colombian Pacific and Caribbean. Mejía-Falla et al. cite four particular collections for their Colombian Bull Shark specimens: the Universidad del Valle (CIRUV, uncataloged, Cali and Buenaventura); the Colección Biológica Marina de Referencia de la Estación Biológica "Henry von Prahl" Parque Nacional Natural Gorgona (PNNG, uncat., Isla Gorgona); the Instituto de Ciencias Naturales de la Universidad Nacional de Colombia (ICN-MHN, uncataloged, Bogotá); and the National Museum of Natural History (USNM00206995, Washington).

Colombia is a bit of a newcomer to elasmobranch research; most survey efforts prior to the 1970s were conducted by foreign entities. However, the Colombian Pacific and Caribbean boast an impressive biodiversity, and may outrival the species richness of adjacent systems; Mejía-Falla et al. specifically contrast the number of Colombian species (88 sharks) with that of the Gulf of Mexico (64 sharks). It should be mentioned, however, that this is a rather unsuitable comparison, as the Gulf of Mexico is a single Atlantic basin, while the Colombian seascape is split between the Caribbean (a likewise Atlantic basin) and the Pacific (an comparatively alien system). Nevertheless, there may be a purpose to the author's disputable example—perhaps to inspire and encourage further investigation into what is, still, a relatively mysterious collection of species.

Mystery is, at least for now, an essential element to shark biodiversity. Leonard Compagno, the renowned taxonomist, reported a global total of 453 shark species in 2005; today—as I write on July 11th, 2016—there are 512 known species of shark. This amazing rate of discovery is a measure of our gap in understanding; how many sharks are *truly* out there? For now, the answer is cryptic…but I sincerely believe that the truth exists in four figures.

2007

Condrictios de la Argentina y Uruguay
**Lista de Trabajo. ProBiota, FCNyM, UNLP, Serie Técnica-
Didáctica, La Plata, Argentina, 11: 1-15
MENNI, R.C. & LUCIFORA, L.O.**

Presentamos aquí una lista actualizada de los condrictios
(tiburones, rayas, quimeras y pez elefante) que han sido citados de la
Argentina y Uruguay, incluyendo las especies de agua dulce de la familia
Potamotrygonidae. Cuando una especie está presente en Argentina o en
Uruguay, se lo indica a continuación de la especie; las demás, son
compartidas. No se indica la distribución de las especies en otras áreas.

Muchas de lase species conocidas de Uruguay terminan allí su
distribución meridional, y ocasionalmente algunas entran a aguas argentinas
y otras podrían hacerlo. Estas presencias ocasionales parecen no ser muy
comunes. La progesiva disminución de las capturas de *Dasyatis violacea*
con la disminucón de temperature hacia el sur ha sido demostrada
(Domingo et al., 2005). En 1981 Menni publicó una lista de las familias de
peces que no sobrepasaban la latitud del Río de La Plata, y de los
condrictios mencionados, Orectolobidae, Ginglymostomatidae y
Rhinopteridae, sólo una especie de la última familia ha sido citada en años
recientes de Uruguay.

Al contrario, especies que han sido citadas de la Argentina, como
Sphyrna tudes de Mar del Plata por Berg (1895) o *Narcine brasiliensis* de la
provincial de Buenos Aires por Lahille (1928), no han sido halladas de
nuevo.

Uruguay is commonly accepted as the southern limit for *Carcharhinus leucas* in the Western Atlantic. However, Menni & Lucifora make note of a remarkable jawset acquired by Chiaramonte in 1998 from a Bull Shark caught in the Argentine province of Buenos Aires. This specimen (now housed in the Argentine Museum of Natural Sciences) may be the Atlantic's southernmost record of a Bull Shark. No other details are given regarding the individual's capture: date, sex, size—all are a mystery.

The water temperatures of Buenos Aires can reach above 20°C from November to March. These summer temperatures of Argentina only slightly overlap with the Bull Shark's minimal preference; it's thus no surprise that the species is rare south of Uruguay. Menni & Lucifora report ten other Uruguayan *Carcharhinus*: *C. acronotus* (Blacknose Shark); *C. brachyurus* (Bronze Whaler); *C. brevipinna* (Spinner Shark); *C. falciformis* (Silky Shark); *C. isodon* (Finetooth Shark); *C. longimanus* (Oceanic Whitetip Shark); *C. obscurus* (Dusky Shark); *C. plumbeus* (Sandbar Shark); *C. porosus* (Smalltail Shark); and *C. signatus* (Night Shark). Of these, only four—the Bronze Whaler, Oceanic Whitetip, Sandbar and Night sharks— are accounted for as Argentine.

We are by now very familiar with the fact that *Carcharhinus leucas* is a fairly cosmopolitan species. However, let us meditate on this shark's massive distribution in the Western Atlantic—the sheer magnitude of such an expansive presence, from a southern limit in Buenos Aires to a northern limit in Massachusetts. Think about it: a coastal stretch of over 9,000 kilometers—9,000 km of Bull Shark. From south to north, this one species patrols a thousand systems: the Rio Grande; the Amazon Estuary; the Lesser Antilles; the Florida Keys; the Outer Banks; the Chesapeake Bay; Cape Cod... Imagine these, so marvelously distant and varied, and yet, transcended by a single name: *leucas*.

2007

On the Record of the Largest (Giant) Bull Shark
Carcharhinus leucas Caught off Chennai
Marine Fisheries Information Service, Technical and
Extension Series, 191: 28-29
RAJAPACKIAM, S. & GOMATHY, S. & RUDRAMURTHY, N.

1233   On the record of the largest (Giant) Bull Shark
          Carcharhinus leucas caught off Chenai

A giant sized female bull shark *Carcharhinus leucas* measuring 356 cm in total length and 320 kg weight was caught by a gillnet operated at a depth of 50-60 m on 22-06-2005. The fishing was done off Iskapalli, Ramathirtham (Nellore), 200 km north of Chennai and landed at Chennai fishing harbor. Other components of the catch were tuna, rays, sailfish, seer fish and carangids.

This shark is the very same individual caught in 2005: the Chennai Bull. New details enrich our understanding of the events surrounding this giant's capture: on June 22nd, 2005, a vessel was trawling a mechanized gillnet in 50-60m of water off the coast of the Indian state of Andhra Pradesh (specifically, off the village of Ramathirtham in the Nellore district). In recovering the net, the fishermen discovered an immense Bull Shark caught amongst a haul of tuna, rays, sailfish, seer fish and carangids. The fisherman travelled 200 km south to Kasimedu Fisheries Harbour (aka Royapuram Fishing Harbour or Chennai Fishing Harbour) in the city of Chennai, Tamil Nadu, India. There the shark was officially landed, weighed and measured: its superlatives of 356 cm TL and 320 kg in weight secured the title of World's Largest. The Chennai Bull was finally auctioned at the landing center for 32,000 Rupees (equivalent to $477.50 USD).

What does it mean to sell the exemplary *Carcharhinus leucas*? This individual successfully hunted, grew, and evaded capture for perhaps 20 or 25 years: the Chennai Bull dwarfed its contemporaries with 20-plus years of muscle building, 20-plus years of sensory tuning and 20-plus years of predatory perfection. Can you imagine the sheer number of teeth shed by this one shark? Thousands upon thousands of keratin gems, white and polished, now litter the seafloor—the lost artifacts of a titan. You could come across such a treasure in your lifetime…but would you know? Is the Chennai Bull really that different from its brothers? Or does it command a profound respect for its attainment of the maximum? I personally find truth in the latter, as the idea of selling such a behemoth for $478 is, well, unsatisfactory. It's as if the Chennai Bull deserves more for its superlative design—not more money, mind you, but just…more. Human needs, of course, must be (and have been) met…but the expense feels rather melancholy.

2007

An Atlas on the Elasmobranch Fishery Resources of India
CMFRI, Special Publication, 95: 1-253
RAJE, S.G. & SIVAKAMI, S. & RAJ, G.M. & KUMAR, P.P.M.
& RAJU, A. & JOSHI, K.K.

The elasmobranchs represented by sharks, skates (sawfishes, guitar fishes) and rays are an important group of demersal fishes which are exploited for multifarious uses of their various body parts such as the meat, fins, liver, teeth and the hide. While shark fins are considered as a delicacy fetching increased export market, their liver oil is utilized in pharmaceutical industry. Shark teeth is used for ornamental purposes and their hide for a variety of leather products. This increased commercial demand coupled with their characteristic life history pattern including slow growth rate, delayed maturation, long reproductive cycle, low fecundity and long life span and their trans-boundary migration pattern make them susceptible to over fishing. Because of this background, of late there had been a growing international awareness over the conservation and management of the elasmobranch stocks. This in turn necessitates an understanding of resource characteristics and eco-biological features of different species of sharks, skates and rays.

The elasmobranchs production during 2003 & 2004 (Average) in India amounted to 57,713t, contributing to 2.25% of the all India marine fish landings (Anon, 2005). Their average annual landings during 1961-2000 in India though indicated a general increasing trend, was found decreasing over the years at states such as Maharashtra, Kerala and Karnataka (Raje and Joshi, 2003).

There is a very strange distributional discrepancy between this report by Raje et al. and the IUCN Red List database. The former states that *Carcharhinus leucas* is exclusive to India's eastern coast on the Bay of Bengal; the latter affirms the opposite, stating that the species is exclusive to the western coast on the Arabian Sea[6]. To make matters even more intriguing, Raje et al. report that the doppelgänger *Carcharhinus amboinensis* is likewise restricted to the Indian west coast; the IUCN conversely states that this species is native to both coasts. Though it's very difficult to disagree with the IUCN's careful reporting on shark species and their distributional ranges, Raje et al. make a compelling case with the specific affirmation that 200t of *Carcharhinus leucas* were landed on the Indian east coast between 2003 and 2004 (of this total catch, 21t were landed by gill net, while 79t were landed by bottom set gill net).

Raje et al. expand the diet of *Carcharhinus leucas* to specifically include mantis shrimp, sea snails and sea urchins, and furthermore suggest that this predator has a, "Tertiary" role in the food chain. Though there is no assessment on the Bull Shark's regional or seasonal abundance, the authors report that this species is slightly threatened: specifically, "Lower Risk: near threatened (LR/nt)" as of June 30th 2000.

This Atlas makes an important contribution in regards to information on the elasmobranch fisheries of both India and the entire Asian continent. In regards to the latter, a complete reporting of tonnes landed in 2003 is provided for each country:

Indonesia	120670	Malaysia	27948	Iran	15963
China	68612	Japan	24906	Korea	12567
India	63266	Thailand	24724	Maldives	11522
Pakistan	33248	Sri Lanka	21290	Yemen	7250

2008

Landing of Female Bull Shark and Napoleon Wrasse
Fish at Tuticorin
CMFRI Newsletter, 119: 8-9
CMFRI

Landing of female bull shark and Napoleon wrasse fish at Tuticorin

On 18.9.2008, a female bull shark (*Carcharhinus leucas*) measuring 311 cm in length was landed at the Tuticorin Fisheries Harbour by a trawler operating at a depth of 50 m. The fish was weighing approximately 320 kg. On the same day one Napoleon wrasse fish or hump head wrasse (*Chelinus undulates*) which is included in the IUCN red list as endangered was landed by another trawler. The fish was measuring 80 cm in TL. This is the second instance of Napoleon wrasse landing at Tuticorin.

(Tuticorin Research Centre of CMFRI)

These short CMFRI accounts may come across as rather trivial for their brevity, but make no mistake, they are essential to the distributional and morphological record of this species. Indian Bull Sharks seem to attain a considerable size, and this is a good sign of ecological health—especially in the face of an increasingly taxed marine ecosystem. The Tuticorin Bull hails from the South India and Sri Lanka marine ecoregion: a heavily impacted environ, where 25.4% of its coastline is densely populated. Only 0.1% of the South India and Sri Lanka shelf is protected by MPAs, and no Marine Stewardship Council Certified Fishery currently exists. [7] Sadly, 92.0% of the South India and Sri Lanka's coral reefs are at risk, and 69.5% suffer from high-impact coral bleaching.[7]

For a Bull Shark to attain such a large size in the face of these ecological pressures is relatively encouraging; the system may not be 'too far gone', so to speak. However, mounting population pressure and economic incentive (for unsustainable fisheries) pose a big risk to this coastal species' health and stability. Size is, in this case, an assessment of overexploitation; if an ecoregion is harvested too heavily, then larger inhabitants are more likely to be captured, while growing inhabitants are less likely to reach the maximum size. This becomes especially important in the consideration of the infamously slow growth rate of Elasmobranchs, as 'large size' may translate to 'breeding size'; without a chance to grow, species like *Carcharhinus leucas* could fail to attain sexual maturity and thereby replenish themselves ecologically (or economically).

We know this dynamic. We've explored it ad nauseam. It's clear as to what is required: protection from fisheries, pollution and other revolting aspects of human impact. Yet, our sense of stewardship moves very slowly, and it's no mystery as to why: sacrifice. Every protective effort costs something, whether in coinage or opportunity.

2008

Movement and Distribution of Young Bull Sharks
Carcharhinus leucas in a Variable Estuarine Environment
Aquatic Biology, 1: 277-289
HEUPEL, M.R. & SIMPFENDORFER, C.A.

The space utilization and distribution of young (<2 yr old) bull sharks *Carcharhinus leucas* within a 27 km stretch of the Caloosahatchee River estuary in Southwest Florida was examined using an array of acoustic monitors to define influences of environmental variables. A total of 56 young sharks from 3 cohorts (2003, 2004, 2005) were fitted with acoustic tags and monitored for up to 460 d. Sharks did not remain within the estuary continuously, but on average approximately one-third were present at any one time from each cohort. Salinity and freshwater inflow showed greatest influence on shark distribution, with temperature appearing to play a limited role. Although individuals occurred in salinites from 0.1 to 34, electivity analysis indicated that they generally avoided areas with salinity <7 and had an affinity for areas with salnities from 7 to at least 20. There were significant relationships between the mean location of a cohort within the estuary and salinity, with sharks occurring further up river when the river was more saline. These relationships were more pronounced for the youngest sharks, and strength of the relationship decreased with age. Since bull sharks are euryhaline, these results suggest that they may select environmental conditions via movement, possibly to reduce energetic costs associated with osmoregulation.

Like all tidal rivers, the Caloosahatchee 'breathes': in and out, tides come and go, while freshwater flows at varying rates from seeps, rains and hurricanes. By such breath, salinity tumbles back and forth, and so inspires the rhythms of estuarine life. Like an excellent waltzer, *Carcharhinus leucas* responds with gentle intuition: in tandem with its saline partner, it never pushes too far to offset the balance, but listens, waits and replies with precision. In so doing, the two 'breathe' together—up and down, push and pull—in delicate balance. The Bull Shark's position: a confident answer to salinity's beckon.

We have seen this dance from a distance; now, Heupel & Simpfendorfer elucidate the form. Thanks to their spectacular research, we now understand that neonate Bull Sharks require much instruction—that before they can handle the sweeping arcs between 0ppt and 30ppt, they must first practice a gentler pattern between 7ppt and 20ppt. This prerequisite may be a necessary compensation for size: smaller Bull Sharks have a higher surface area to volume ratio than larger individuals. This critical difference in physique is essential to proper osmoregulation: smaller Bull Sharks, "require greater amounts of energy to regulate water and ions if exposed to varying salinities, especially given the importance of urea in the osmoregulatory strategy of sharks."

The incapability of young Bulls to fully regulate their urea is what likely prohibits them from exiting the mild, 'iso-osmotic' conditions of the estuary. Only with age (around 18 months) do they embolden: increased size translates to a decreased surface area to volume ratio, and a keener grasp on the tact of osmoregulation. Until that time, the youngest Bulls must remain in the gentlest position, and conserve their energy; by reducing osmoregulatory costs, young *Carcharhinus leucas* (presumably) allocate more energy to growth—to better footing for the impending grander waltz.

2008

Systematic, Paleoecologic and Paleobiogeographic
Analysis of the Plio-Pleistocene
Mediterranean Elasmobranch Fauna
Atti Societa Toscana Scienze Naturali, Serie A, 113: 81-88
MARSILI, S.

A synthesis of the diversity of the Plio-Pleistocene elasmobranch fauna of the Mediterranean Sea is provided. The data deriving from the revision of Italian fossil elasmobranch teeth, and/or isolated elasmobranch teeth records of the Mediterranean area, are used to compile a database including all the Plio-Pleistocene elasmobranch taxa. 72 elasmobranch species included in 51 genera, 26 families, and 11 orders have been recognized. The revision of several fossil teeth from the Lower-Middle Pliocene deposits of Italy provides the first Pliocene record of the extinct *Isurus xiphodon* in the Mediterranean. Moreover, the Pliocene occurrence of *Carcharodon megalodon* into this basin is supported by six large teeth recorded in the Italian fossil teeth elasmobranch collections. The qualitative and quantitative analysis of the database confirms a decreasing tend in elasmobranch taxonomic diversity across the Plio-Pleistocene. The paleoecological and paleobiogeographical evolutionary patterns involved in the establishment of the present epi-mesopelagic, and deep-water Mediterranean elasmobranch fauna are also discussed.

Key words – Pliocene, Pleistocene, Elasmobranchii, shark, fish fauna, paleoecology, paleogeography, climate change, Mediterranean Sea.

Long ago, a terrific cataclysm altered the face of the Miocene Mediterranean, and forever changed the destiny of its elasmobranchs: the Messinian Salinity Crisis. During this period within the Upper Miocene (about 6.0 to 5.6 Ma), the narrow western passage that would one day become the Strait of Gibraltar closed. This severance of the Atlantic from the Mediterranean proved to be disastrous: with no oceanic exchange (in the face of a drier climate), the basin began to evaporate. As water escaped into the atmosphere, salinity drastically increased, and soon the entire Mediterranean became hypersaline. With this consequence, desiccation was inevitable: the Messinian Mediterranean was nearly burned away. Some water remained—but in pockets miles below sea level.

Could the Ancient Bull have endured such chaos? Marsili indicates that, "most of the Miocene biota survived in refugee areas or disappeared from the Mediterranean across the Upper Miocene-Lower Pliocene boundary". Given that *Carcharhinus leucas* fossils have been found in Miocene Mediterranean strata, the answer could (amazingly) be, 'yes'. Conversely, Marsili also asserts that *Carcharhinus leucas* may have arrived with the Zanclean flood: an early Pliocene (5.3 Ma) resurgence of the Mediterranean Sea, in which the Atlantic poured back over the once-closed (ancient) Strait of Gibraltar. This miraculous replenishment of the basin, "allowed the establishment of a rich marine biota, with a permanent Atlantic affinity." Numerous subtropical sharks once foreign to the Mediterranean (such as *Galeocerdo cuvier*, *Carcharhinus perezi* and *Rhizoprionodon* aff. *acutus*) emerged during this period of life and revitalization; *Carcharhinus leucas* may have been among them, and thus established its presence in the early-middle Pliocene.

Whatever the origin of the Ancient Bull, it disappeared with the cold of the Pliocene—what Marsili calls a, "climatic deterioration".

2008

*Discrimination of Shark Species by Simple
PCR of 5S rDNA Repeats*
Genetics and Molecular Biology, 31, 1 (suppl): 361-365
PINHAL, D. & GADIG, O.B.F. & WASKO, A.P. & OLIVEIRA,
C. & RON, E. & FORESTI, F. & MARTINS, C.

Sharks are suffering from intensive exploitation by worldwide
fisheries leading to a severe decline in several populations in the last
decades. The lack of biological data on a species-specific basis, associated
with a k-strategist life history make it difficult to correctly manage and
conserve these animals. The aim of the present study was to develop a
DNA-based procedure to discriminate shark species by means of a rapid,
low cost and easily applicable PCR analysis based on 5S rDNA repeat units
amplification, in order to contribute conservation management of these
animals. The generated agarose electrophoresis band patterns allowed to
unequivocally distinguish eight shark species. The data showed for the first
time that a simple PCR is able to discriminate elasmobranch species. The
described 5S rDNA PCR approach generated species-specific genetic
markers that should find broad application in fishery management and trade
of sharks and their subproducts.

Key words: Chondrichthyes, PCR, species identification, 5S rDNA, sharks.

The progress of science is akin to a coral. Each breakthrough dazzles like a living polyp, and actively contributes to the luminous complex. One day, each life will wither, but only to create a skeletal base for the next great discovery: in this way, knowledge grows.

Pinhal et al.'s contribution to elasmobranch genetic identification is exemplary of this brilliant dynamic. In reference to a familiar skeleton (multiplex PCR assay using mitochondrial cytochrome b—a method which once was claimed to be the most economical for genetic ID), Pinhal et al. construct a newer, more efficient technique: a combination of PCR amplification of 5S ribosomal DNA repeats (5S rDNA) and agarose gel electrophoresis analysis. The authors present this new approach as, "a simple routine and low cost methodology to achieve shark species identification." There are many advantages to utilizing the 5S rDNA array, and they derive primarily from its structure: a constitution of, "multiple copies of a highly conserved 120 base pairs (bp) coding sequence, separated from each other by a variable non-transcribed spacer (NTS)".

With the primers Cart5S1F and Cart5S1R, Pinhal et al. were able to successfully amplify the 5S rDNA repeats of eight shark species and compare their distinct agarose gel fragment patterns. *Carcharhinus leucas* exhibited a unique combination of four bands (230, 450, 500 and 540 bp). The Dusky Shark, *Carcharhinus obscurus*, shared three of these bands, but lacked the Bull Shark's exclusive addition of 230 bp. Though the 5S rRNA gene is highly conserved in itself, there is (as evident by each species' distinct gel fragment patterns) a, "high variability in the 5S rDNA genomic architecture." Pinhal et al. propose that it is the NTS organization of each species that varies immensely, and that the NTS regions themselves, "seem to be subject to intense evolution".

2009

Estimating Population Parameters for Bull Sharks at a Feeding Site in a Marine Protected Area in Fiji Abstract
In: Progrtamm and Abstracts, 13th EEA Conerence 2009,
Palma de Mallorca, Spain
(ed.Morey, G. & Yuste, L. and Pons, G.X.): 59
BAENSCH, H. & BRUNNSCHWEILER, J.M.

Marine megafauna attracts increasing attention in ocean conservation planning and threatened predators, such as sharks, are used to promote marine protected areas. Such sites attract divers which has led to the growth in the popularity of marine wildlife watching as a marine tourism activity, but also provide a good opportunity to estimate fish abundance and life-history parameters. We assessed biological and population parameters of the bull shark, *Carcharhinus leucas*, at a stationary feeding site in the Shark Reef Marine Reserve, Fiji, through direct underwater observation between 2003 and 2008. The number of bull sharks decreases over the course of a calendar year with fewest sightings between September and November. Based on body conditions of females, such as pregnancy and mating scars, we hypothesize that bull shark absence from the site towards the end of a calenadar year is related to reproductive activity. Peak abundances at the Shark Reef Marine Reserve occur in the first half of the year with most sightings in February. The female:male sex ration is 3.64:1. Based on natural distinctive marks for individual recognition, direct observation led to the identification of 54 individuals (7 males, 47 females). Of these, 29 (5 males, 24 females) were reliably identified in multiple years with 89.7% sharks observed over at least 2 years indicating site-fidelity.

To address *Carcharhinus leucas* as a species is to apply a collective; though we have an archetypical image of the essential characters that define *Carcharhinus leucas*, there exists, in truth, a compelling diversity of individual deviations. Like the wonderful diversity that is human individuality, so too is the complexity of intraspecific variance. Shark Reef Marine Reserve and its associated SCUBA program, Beqa Adventure Divers, do a marvelous job recording the unique marks, scars and behaviors of the local populace of Bulls. Each (readily identifiable) individual is given a name, and by extension, a profile: the name serves as the backbone of identity, and it is to this that scientists and divemasters apply their behavioral observations to construct (in essence) a personality.

Isn't it incredible that each Bull Shark possesses a unique set of genes and experiences? Isn't it marvelous that, for each shark, these should combine to form a unique perception? It's almost overwhelming to comprehend such diversity. Like humans, Bull Sharks can be shy or bold, gentle or agitated—their temperaments vary in relation to evolutionary lineage and ecological influence. Consider, then: which positive impacts, negative pains and biological precursors have molded each shark's perspective?

Here is a select list of the colorful Bulls from Beqa Lagoon[8]:

Name	Marks	Personality
Kinky	'Kink' in front of 1st dorsal	Fussy eater; spits out food
Stumpy	Top of caudal fin 'cut off'	Self-assured, relaxed; "sweet"
Cilla	Severed right pelvic fin	Very timid
Jaws	Jaw hangs open on left side	Very gentle and easy going
Whitenose	White patch on tip of nose	A bit edgy; may affect others
Blackbeard	Hook in R corner of mouth	Very cautious (despite size)

2009

The Seasonal Importance of Small Coastal Sharks and Rays in the Artisanal Elasmobranch Fishery of Sinaloa, Mexico
Pan-American Journal of Aquatic Sciences, 4 (4): 513-531
BIZZARRO, J.J. & SMITH, W.D. & CASTILLO-GÉNIZ, J.L. & OCAMPO-TORRES, A. & MÁRQUEZ-FARÍAS, J.F. & HUETER, R.E.

Seasonal surveys were conducted during 1998-1999 in Sinaloa, Mexico to determine the extent and activites of the artisanal elasmobranch fishery operating in the southeastern Gulf of California. Twenty-eight fishing sites were documented, the majority of which (78.6%) targeted elasmobranchs during some part of the year. Sharks numerically dominated sampled landings (65.0%, $n = 2390$), and catch rates exceeded those of rays during autumn — spring. The scalloped hammerhead, *Sphyrna lewini*, was the primary fishery target during these seasons, with most landings composed of early life stages. During summer, rays, especially *Rhinoptera steindachneri*, were numerically dominant (87.7%). Large sharks were of comparably minor importance in the artisanal fishery during all seasons. Catch composition was similar between spring and winter ($SIM_{obs} = 0.393$, $SIM_{exp} = 0.415$; $P = 0.25$), largely because the fishery mainly targeted "cazón" (sharks ≤ 1.5 m total length) during this period (e.g., *S. lewini*, *Rhizoprionodon longurio*). Small size classes of large sharks and a wide size range of coastal sharks and rays were primarily observed. In addition, size composition of *S. lewini* and to a lesser extent, *R. longurio* decreased significantly between historic and contemporary landings. Local populations of these species should therefore be closely monitored.

Only two Bull Sharks—a 123 cm female and 182 cm male—were captured during Bizzarro et al.'s survey (comprising only 0.1% of the total catch and 0.4% of the summer catch). Though *Carcharhinus leucas* appears to be one of the least common "tiburón" (large sharks over 1.5m TL) in the Sinaloa shark fishery, it is nonetheless one of the many sharks at risk of Sinaloa's exploitative practices. Sinaloa is a national leader in elasmobranch production and constitutes a large percentage of Mexico's total catch (16.5% in 2006). Tuna, sardines and shrimp are the state's most important fisheries resources, and altogether form the backbone of Mexico's largest (or second largest) fishing industry (Sinaloa accounted for 17.3% of Mexican landings and 20.7% of revenue in 2006). Elasmobranchs—though not a primary target—form a significant fisheries component at risk of overexploitation. Though the directed harvest is destructive, bycatch (chiefly in relation to Sinaloa's shrimping industry) appears to have the severest impact.

The biggest concern with the Sinaloa shrimping industry is the fact that trawling often takes place in the nursery areas of sharks and rays; this troubling operation may, "represent a considerable source of mortality for early life stages." Indeed, the average sizes of *Sphyrna lewini* (Scalloped Hammerhead) and *Rhizoprionodon longurio* (Pacific Sharpnose Shark)— two of Sinaloa's most important "cazón" (small sharks under 1.5 m in TL)—have decreased: as reported by the authors, "contemporary landings of both species reflect a considerable reduction in size composition when compared to those of 1999." The decline of *Rhizoprionodon longurio* is particularly disturbing, as it is (among the sharks) a fast-growing, quick-to-reproduce species—one of the most resilient sharks in the face of fisheries pressure. Large, slow-growing species like *Carcharhinus leucas* could not well endure such a force as to truncate the sturdy, steadfast Sharpnose.

2009

Oceans Apart? Short-term Movements and Behaviour of
Adult Bull Sharks in Atlantic and Pacific Oceans
Determined from Pop-off Satellite Archival Tagging
In: Progrtamm and Abstracts, 13th EEA Conference 2009,
Palma de Mallorca, Spain
(ed. Morey, G. & Yuste, L. and Pons, G.X.): 61
BRUNNSCHWEILER, J.M. & QUEIROZ, N. & SIMS, D.W.

Satellite telemetry can provide insight into movement patterns and behaviour of coral-reef or coastal shark species such as the bull shark, *Carcharhinus leucas*. In this study, we monitored adult bull sharks with electronic tags to gain insight into horizontal movements, including the species' potential for long-distance migration, at two aggregation sites in the Atlantic and Pacific Oceans and evaluated vertical behaviour in terms of time spent at depth and temperature. In both locations, bull sharks showed some fidelity to specific coastal areas with only limited horizontal movements away from the tagging sites after tag attachment. In the Bahamas, bull sharks were detected mostly in the upper 20 m of the water column in water 25-26° C, whereas sh arks tagged in Fiji spent most of their time below 20 m in water usually above 26° C. The results show that bull sharks are important parts of coastal ecosystems and underpin the need for international and national cooperation when devising conservation strategies for this species.

Fidelity is a recurring quality when it comes to *C. leucas*. Previous studies on the Fijian Bulls of Beqa Lagoon have also reported high site fidelity, while research on the Floridian Bulls (adjacent to the Bahamian system) likewise indicate a rather localized distribution. Though many populations of Bull Sharks do indeed migrate, their journeys are not as famed (or extensive) as those of the White, Tiger and Blue sharks—the classic nomads. Instead, *C. leucas* appears to be more of a homebody—a species that, while venturous, prefers the patrol of acquainted haunts.

As in the romantic sense, ecological fidelity forms the bedrock of intimacy: in remaining loyal to a specific place, *Carcharhinus leucas* enriches a mutual sense of familiarity. Immersion into the Fijian or Bahamian system renders the Bull Shark as *characteristically* Fijian or Bahamian: the experiences of one place will not match the experiences of another, as each will exert a unique but powerful ecological impact on its respective Bulls. For example, a Bahamian *Carcharhinus leucas* exhibiting high site fidelity will largely draw from a uniquely Bahamian experience: specific resources, temperatures, fishing pressures and habitat availability will all combine into a specific impact—a radiation, if you will, that alters *Carcharhinus leucas* into a 'Bahamian state'.

Likewise, 'Fijian Bulls' exhibiting high site fidelity become uniquely molded into the Fijian seascape, and thus exist as an integral part of the Fijian community. This latter point is especially important, as site fidelity establishes a healthy sense of dependence and expectation—the *expectation* that *Carcharhinus leucas* exists as a Fijian component. It is from this stable platform that scientists and ecotourists alike can construct a robust research project or dive operation; Beqa Adventure Divers is an excellent example of the economic benefits of site fidelity, as the operation could not exist without *expected* Bulls.

2009

**Physical Factors Influencing the Distribution of a Top
Predator in a Subtropical Oligotrophic Estuary**
Limnology and Oceanography, 54 (2): 472-482
HEITHAUS, M.R. & DELIUS, B.K. & WIRSING, A.J. &
DUNPHY-DALY, M.M.

We used longline fishing to determine the effects of distance from
the ocean, season, and short-term variation in abiotic conditions on the
abundance of juvenile bull sharks (*Carcharhinus leucas*) in an estuary of
the Florida Everglades, U.S.A. Logistic regression revealed that young-of-
the-year sharks were concentrated at a protected site 20 km upstream and
were present in greater abundance when dissolved oxygen (DO) levels were
high. For older juvenile sharks (age 1+), DO levels had the greatest
influence on catch probabilities followed by distance from the ocean; they
were most likely to be caught at sites with >3.5 mg L^{-1} DO and on the main
branch of the river 20 km upstream. Salinity had a relatively small effect on
catch rates and there were no seasonal shifts in shark distribution. Our
results highlight the importance of considering DO as a possible driver of
top predator distribution in estuaries, even in the absence of hypoxia. In
Everglades estuaries hydrological drivers that affect DO levels (e.g.,
groundwater discharge, modification of primary productivity through
nutrient fluxes) will be important in determining shark distributions, and the
effects of planned ecosystem restoration efforts on bull sharks will not
simply be mediated by changing salinity regimes and the location of the
oligohaline zone.

Three primary colors once painted the distribution of *Carcharhinus leucas* in Florida: temperature, site and salinity. Now, with Heithaus et al., we have a fourth color to consider—an exciting addition that adds sharpness and depth to our understanding of the Bull Shark's ecological portrait: dissolved oxygen (DO). This element appears to have the highest distributional influence on *Carchahrinus leucas* in the Shark River Estuary—even more powerful that that of site and salinity. Indeed, Heithaus et al. question the impact of salinity as asserted by older claims—specifically those of authors like Heupel & Simpfendorfer (who have past stated that juvenile Bull Sharks prefer an optimal salinity range between 7ppt and 20ppt). Half of this study's young of the year (YOY) were caught outside a comparable range of 7.05ppt and 17.45ppt, leading the authors to believe that, "salinity was not a significant predictor of YOY catches".

As dissolved oxygen has a tremendous impact on the distribution of Bull Sharks, so too does it have on the Shark River's entire trophic structure. Low levels of DO (specifically below 3.5 mg/L) largely deter Bull Sharks, and thus create an opportunity for potential prey items (able to resist low DO) to escape the jaws of *C. leucas*. However, DO-independent predators (such as the American Alligator) can exploit this opportunity, and replace the Bull Shark as the system's top consumer; as such, low DO is less likely to protect prey and, "more likely to mediate the predator type in a particular location".

This is incredibly important to understand in light of the fact that Shark River—located in Everglades National Park—is one the last, "undeveloped nursery areas for bull sharks in the southeast United States." Climate change and restoration efforts affecting the water flow of the Everglades may alter DO concentrations, which in turn could affect the availability of prey items for *C. leucas* at its most vulnerable stage of life.

2009

Overview of the U.S. East Coast Bottom
Longline Shark Fishery, 1994-2003
Marine Fisheries Review, 71 (1): 23-38
MORGAN, A. & COOPER, P.W. & CURTIS, T.H. &
BURGESS, G.H.

The U.S. Atlantic coast and Gulf of Mexico commercial shark fisheries have greatly expanded over the last 30 years, yet fishery managers still lack much of the key information required to accurately assess many shark stocks. Fishery observer programs are one tool that can be utilized to acquire this information. The Commercial Shark Fishery Observer Program monitors the U.S. Atlantic coast and Gulf of Mexico commercial bottom longline (BLL) large coastal shark fishery. Data gathered by observers were summarized for the 10-year period, 1994 to 2003. A total of 1,165 BLL sets were observed aboard 96 vessels, with observers spending a total of 1,509 days at sea. Observers recorded data regarding the fishing gear and methods used, species composition, disposition of the catch, mortality rates, catch per unit effort (sharks per 10,000 hook hours), and bycatch of this fishery. Fishing practices, species composition, and bycatch varied between regions, while catch rates, mortality rates, and catch disposition varied greatly between species.

Carcharhinus leucas is still an uncommon species in the American shark fisheries of the Mid Atlantic Bight, Southeastern Atlantic and Eastern Gulf of Mexico; according to the Commercial Shark Fishery Observer Program (CSFOP), Bull Sharks comprised only 1.2% of the total catch from 1994 to 2003. Though *Carcharhinus leucas* is not a traditionally targeted species, the majority of captured individuals were landed (94.6%); only 1.2% of the catch was released, while the remainder either escaped (2.8%) or endured other fates (resulting in a total mortality rate of 95.7%). At-vessel mortality was low for *Carcharhinus leucas*; 69.9% of the catch was alive upon retrieval (making this species one of the hardiest in the face of longline capture stress). Morgan et al. speculate that Bull Sharks, "possibly suffer lower at-vessel mortality rates because they are not obligate ram ventilators and may not become greatly stressed when hooked." Indeed, the highest at-vessel mortality rates were observed for the hammerheads, which are classic obligate ram ventilators (these species require constant movement for oxygenation, and the spatial limitations of the longline presumably interfere with this vital process).

Carcharhinus leucas demonstrated an increase in Catch Per Unit Effort (CPUE) over the study period ($r^2 = 0.28$). Peak catch rates occurred in 1995, followed by, "2 years of declining CPUE, and then a subsequent rise to a new peak in 1999". Though CPUE is a commonly used indicator of stock size (and its fluctuations), Morgan et al. introduce a number of critical concerns regarding CPUE's accuracy as a stock assessment—the most pertinent of which being the idea that CPUE increases for the Bull Shark (and others) maybe, "a result of fishermen becoming more adept at targeting and catching specific species." A shallow interpretation of CPUE alone would have assumed that *Carcharhinus leucas* stocks were increasing; unfortunately, such would not be the disheartening truth.

2009

Movement Patterns and Water Quality Preferences of
Juvenile Bull Sharks (Carcharhinus leucas) in a
Florida Estuary
Environmental Biology of Fishes, 84 (4): 361-373

**ORTEGA, L.A. & HEUPEL, M.R. & VAN BEYNEN, P. &
MOTTA, P.J.**

Acoustic telemetry was used to examine the size of daily activity space, small-scale movement patterns, and water quality preferences of juvenile bull sharks in the Caloosahatchee River, Florida. Movement pattern analysis included rate of movement, swimming depth, linearity, direction, tidal influence, diel pattern, and correlation with environmental variables. Manual tacking occurred before and after a large freshwater influx which divided the sharks into two groups based on movement patterns. The first group displayed increased rate of movement, distance traveled, and space utilization at night, and movements correlated with salinity, temperature, and dissolved oxygen. The second group had an increased rate of movement, distance traveled, and space utilization during the day, and movements correlated with temperature, dissolved oxygen, turbidity and pH. These juvenile bull sharks displayed distinct diel movement patterns that were influenced by physical factors, which may account for the distribution of this top-level predator in the Caloosahatchee River.

Keywords *Carcharhinus leucas* Habitat use Activity space Movement patterns Acoustic telemetry Manual tracking

Details as crystallized by Ortega et al. proudly demonstrate the estuary's command of its citizenry: tidal, temporal and trophic forces each wield an exceptional say on the Bull Shark's location, and find themselves further amplified by the Caloosahatchee's shape and proximity to Lake Okeechobee. Anthropogenic influence (such as the artificial release of freshwater from the Lake) has a profound impact on this intricate dynamic, and elicits the special concern of Ortega et al. (who understand the Caloosahatchee's importance as a nursery ground for *Carcharhinus leucas*).

Tidal fluctuation is the ecological cornerstone of many behaviors. As the tide falls, *Carcharhinus leucas* travels downriver and swims at a greater depth in a more linear pattern. Linearity is in itself linked to a faster rate of movement; the preference for depth, meanwhile, could be, "a means of remaining in a desired environmental regime". Specifically, juvenile Bull Sharks moving with the falling tide may prefer the well-mixed (and saltier) water near the bottom, as it could be less of an osmoregulatory stressor than the poorly mixed (and fresher) water near the surface. When the tide rises, *Carcharhinus leucas* travels upriver and swims at a shallower depth in a more random pattern. Randomness is in itself linked to a slower rate of movement, but Ortega et al. assert that their juvenile Bull Sharks, "moved at an elevated speed as they traveled upriver".

In tandem with the tide is time: juvenile Bull Sharks prefer deeper water during the day, but swim closer to the surface at night. This temporal preference may be sincere on the part of *Carcharhinus leucas*; alternatively, it could be a response to diel variations in prey distribution (i.e., teleosts preferring the surface at night). Food is, after all, a compelling dictation: it's quite unsurprising that many of the Bull Shark's locational preferences (especially in relation to tidal movement) are in synch with those of its prey.

2009

Identification and Phylogenetic Inferences on Stocks of Sharks Affected by the Fishing Industry off the Northern Coast of Brazil
Genetics and Molecular Biology, 32 (2): 405-413
RODRIGUES-FILHO, L.F.S. & ROCHA, T.C. & REGO, P.S. &
SCHNEIDER, H. & SAMPAIO, I. & VALLINOTO, M.

The ongoing decline in abundance and diversity of shark stocks, primarily due to uncontrolled fishery exploitation, is a worldwide problem. An additional problem for the development of conservation and management programmes is the identification of species diversity within a given area, given the morphological similarities among shark species, and the typical disembarkation of processed carcasses which are almost impossible to differentiate. The main aim of the present study was to identify those shark species being exploited off northern Brazil, by using the 12S-16S molecular marker. For this, DNA sequences were obtained from 122 specimens collected on the docks and the fish market in Bragança, in the Brazilian state of Pará. We identified at least 11 species. Three-quarters of the specimens collected were either *Carcharhinus porosus* or *Rhizoprionodon* sp, while a notable absence was the daggernose shark, *Isogomphodon oxyrhyncus*, previously one of the most common species in local catches. The study emphasizes the value of molecular techniques for the identification of cryptic shark species, and the potential of the 12S-16S marker as a tool for phylogenetic inferences in a study of elamsobranchs.

The use of genetic markers as an effective (and convenient) means of species identification is by now a very familiar topic. However, Rodrigues-Filho et al. present an exciting novelty: a phylogenetic assessment of the Carcharhiniformes based on a 1,380 base-pair sequence from the 12S-16S region of mitochondrial DNA. Though this is by no means a "final" verdict on the relationships between *Carcharhinus leucas* and the other Carcharhiniformes, the authors' phylogram is nonetheless a significant achievement. It appears that the 'ridgeless' *Carcharhinus* are very likely monophyletic: in corroboration with previous studies, Rodrigues-Filho et al. place *Carcharhinus leucas* most closely with *C. brevipinna*, and furthermore group these two into a monophyletic clade shared by *C. acronotus*, *C. isodon*, *C. porosus* and (most intruigingly) *Isogomphodon oxyrhynchus*.

The following is a table of genetic divergence values (uncorrected "p" distance), which indicate the "genetic distance" between *C. leucas* and the other Carcharhiniformes (as sampled by the authors). The values (%) are arranged by distance (0.00% = closest).

Species Name	"p" (%)	Species Name	"p" (%)
Carcharhinus leucas	**0.13**	*Mustelus canis*	**6.96**
Carcharhinus acronotus	**3.66**	*Rhizoprionodon* spp.	**7.30**
Carcharhinus falciformis	**3.70**	*Mustelus norrisi*	**8.06**
Isogomphodon oxyrhynchus	**4.13**	*Galeocerdo cuvier*	**8.63**
Carcharhinus altimus	**4.13**	*Sphyrna tiburo*	**8.83**
Carcharhinus perezi	**4.22**	*Sphyrna tudes*	**8.88**
Carcharhinus plumbeus	**4.45**	*Sphyrna mokarran*	**9.33**
Carcharhinus limbatus	**4.95**	*Sphyrna lewini*	**9.40**
Carcharhinus porosus	**4.99**	*Sphyrna* sp.	**10.00**

2009

Carcharhinus leucas

In: IUCN 2012. IUCN Red List of Threatened Species.
Version 2012.2. <www.iucnredlist.org>
SIMPFENDORFER, C.A. & BURGESS, G.H.

Assessment Information [top]

Red List Category & Criteria:	Near Threatened ver 3.1
Year Published:	2009
Date Assessed:	2005-10-01
Annotations:	Needs updating
Assessor(s):	Simpfendorfer, C. & Burgess, G.H.
Reviewer(s):	Musick, J.A. & Fowler, S.L. (Shark Red List Authority)

Justification:

This assessment is based on the information published in the 2005 sharks status survey (Fowler *et al.* 2005).

The Bull Shark (*Carcharhinus leucas*) is a common tropical and subtropical species that occurs in marine, estuarine and freshwater. It is the only species of shark that can exist for long periods in freshwater and penetrates long distances up large rivers. It is caught in fisheries throughout its range, but it is rarely a target species. Its occurrence in estuarine and freshwater areas makes it more vulnerable to human impacts and habitat modification.

This assessment from the IUCN Red List is an excellent species summation of *Carcharhinus leucas*; however, a few regrettable mistakes poison the account's accuracy as a whole. Firstly, the authors incorrectly state that *Carcharhinus leucas* is, "the only species of shark that can exist for long periods in freshwater". Multiple sharks of the genus *Glyphis* extensively utilize estuarine and freshwater habitat (hence their common collective of 'River Sharks'); this is not new information and should have been taken into account before the inclusion of such an exemplary claim. Secondly, multiple American states have been wrongfully excluded from the Native Countries Occurrence section (TX, AL, GA, SC, NC, VA, MD, DE, NJ, NY, RI, CT, MA); simultaneously, WI, MN, and IA have all been shockingly included (possibly as a regrettably serious interpretation of a classic hoax—akin to the idea of a Bull Shark in the Great Lakes).

Such woefully inaccurate claims are incredibly dangerous to preserve within a body as globally influential as the IUCN. Governments, businesses and others consult the IUCN as a global conservation leader; for an IUCN assessment to have inaccuracies (however small) is to weaken the assessment's integrity as a whole. This can damage confidence in the IUCN as a conservation authority; one who may be invested in the protection of *C. leucas* could easily wonder, "If the IUCN is incorrect about the Bull Shark's range, then *what else* is the IUCN incorrect about the Bull Shark?"

Concerns aside, the majority of the IUCN's assessment on *Carcharhinus leucas* corroborates with our prior knowledge of the species. Specific vital stats include Smith et al.'s (1998) estimated natural mortality (0.166/yr., based on a maximum age of 27 years) and rebound potential (0.027-0.039/yr.). Growth rates from both Thorton & Lacy (1982) and Branstetter & Stiles (1987) are also included (spanning from 18 cm/yr. to 4 cm/yr.).

2010

Marine Fishes of the Caribbean Coast of Lower Central
America An Illustrated Guide
Revista de Biología Tropical, 58 (Suppl. 2): 232pp
BUSSING, W. & LÓPEZ, M.I.

INTRODUCTION

The present manual is the third in a series of illustrated guides to the identification of Costa Rican marine fishes. We include in this volume all species found in coastal waters along the Caribbean side of Costa Rica: reef dwellers, demersal or bottom dwellers, inshore pelagic species, and those euryhaline species usually found in estuaries and mangrove swamps. All species presently deposited in the Museo de Zoología of the University of Costa Rica (UCR) form the basis for this volume. We have been fortunate to have obtained many figures and photographs from a large variety of sources. The diversity of figures from different sources makes it especially necessary to focus on the diagnostic characters in the text. We consider this manual to be useful as a field guide and in the preliminary sorting of specimens in the laboratory. Numerous other publications should be consulted to make definite identifications, especially the 2002 FAO Guide.

For many years we had realized the need for a thoroughly illustrated guide to Caribbean fishes found specifically along the coastline of Costa Rica. When we initiated our studies of Costan Rican fishes in the sixties, the three volume set of Seth Meek and Samuel Hildebrand (1923-1928) was our principal identification guide to the marine fishes of this region. Certainly most of the common fishes were included in their treatise.

Very little is said about *Carcharhinus leucas* in Costa Rica: only the essential characteristics are provided, namely, "Short, blunt snout; heavy and deep-bodied." It thus may be prudent to forgo an overly detailed analysis and instead take a moment to reflect: what has happened with *Carcharhinus leucas* at the dawn of the 21st Century?

Our genetic understanding has sharpened. We now have a handful of molecular markers for easy species identification, and furthermore find ourselves in an exciting position to explore the Bull Shark's phylogeny in detail. It appears that *Carcharhinus leucas* is monophyletic with the other *Carcharhinus* lacking an interdorsal ridge. This is both odd and intriguing, as *Carcharhinus leucas* boasts the largest size and heaviest physique of this 'ridgeless' group (recall that the other *Carcharhinus* are much smaller and slimmer). If this group is truly monophyletic, then what evolutionary pressures have molded each shape? How did the Bull Shark attain such a uniquely powerful form?

Fisheries and life history data on *Carcharhinus leucas* appear to have been perfected within the new millennium's development; we now have a keener sense of the Bull Shark's vital parameters (such as variable growth rates, natural mortality, rebound potential, etc.) and a greater understanding of its fisheries role in many countries (especially the United States, though *Carcharhinus leucas* exerts little impact on both commercial and recreational operations). Interest in the Bull Shark's abiotic preferences (such as salinity, dissolved oxygen and temperature) has grown within the past decade, yet it feels as if we have just begun to understand this species' precise environmental regime (as critical research on the subject appears to conflict). Of final note is the rising awareness of habitat alteration and its impact on *Carcharhinus leucas*; before 2010, it existed in the form of power plants, dams and beach mesh. What is to come after?

2010

Distributions of Sharks across a Continental Shelf in the
Northern Gulf of Mexico
Marine and Coastal Fisheries: Dynamics, Management, and
Ecosystem Science, 2 (1): 440-450
DRYMON, J.M. & POWERS, S.P. & DINDO, J. &
DZWONKOWSKI, B. & HENWOOD, T.A.

Declines in shark populations have sparked researchers and fishery
managers to investigate more prudent approaches to the conservation of
these fish. As managers strive to improve data collection for stock
assessment, fisheries-independent surveys have expanded to include data-
deficient areas such as coastal regions. To that end, a catch series from a
nearshore survey off Alabama was combined with data from a concurrent
offshore survey with identical methodology to examine the depth use of
sharks across the continental shelf (2-366 m). The combined data set
contained 22 species of sharks collected from 1995 to 2008: 21 species in
the offshore data set (1995-2008) and 12 species in the nearshore data set
(2006-2008). Depth was a significant factor determining species'
distributions, primarily for Atlantic sharpnose *Rhizoprionodon terraenovae*,
blacknose *C. acronotus*, and blacktip *C. limbatus* sharks. Blacknose sharks
had the highest catch per unit effort (CPUE) in the middepth stratum (10-30
m), blacktip sharks had consistently higher CPUE in the shallow depth
stratum (<10 m), and Atlantic sharpnose sharks showed high abundance
throughout both the shallow and middepth strata. Length frequency and sex
ratio analyses suggest that Atlantic sharpnose and blacknose sharks are
using waters greater than 30 m deep for parturition, whereas adult blacktip
sharks are probably using shallow waters for partuition.

Drymon et al. present a critical case against the distributional inaccuracies of previous shark fisheries surveys. Such surveys (be they of Gulf Coast or Atlantic origin) have certainly enriched our understanding of American species composition, but have largely failed to do so with respect to depth—a key ecological preference. With variance in depth comes variance in temperature and salinity; the ecological regime of deeper water is saltier, colder and typically more stable than that of the shallows. Certain species (or individuals within a species) may prefer this regime to the surface; thus, a purely surface-level understanding of distribution (as seen in previous studies) is inadequate.

Drymon et al. remedy this issue by crystallizing depth's effect on the intra-specific distributions of 22 shark species; of these 22 sharks, the Blacknose, Blacktip and Atlantic Sharpnose yielded the most robust findings. Blacknose and Atlantic Sharpnose Sharks demonstrated the highest CPUE at middle depths (10-29.9 m), while Blacktip Sharks yielded a significantly higher CPUE in the shallows (0-9.9m). Sex ratios also varied significantly for each of these species in respect to depth: Atlantic Sharpnose and Blacknose Sharks showed a prominent bias towards females in deeper water, while Blacktip Sharks conversely demonstrated a strong female bias in the shallows. Findings like these emphasize the importance of depth as an ecological determinant of distribution, and demand that future surveys accommodate for such a critical management factor.

Bull Sharks exhibited the highest CPUE in shallow water (CPUE = 0.33), and were captured, "at a rate of four times higher in shallow waters than in deep waters". The low deepwater catch rate (CPUE = 0.08) suggests, "a coastal life history in the northern Gulf of Mexico." This is in juxtaposition with the Blacknose, Blacktip and Sharpnose Sharks, which each experience a, "sexual segregation of life history phases across depth."

2010

Systematics, Paleobiology, and Paleoecology of Late
Miocene Sharks (Elasmobranchii, Selachii) from Panama:
Integration of Research and Education
Master Thesis: Gainesville: University of Florida. 131 p.
PIMIENTO, C.

The late Miocene Gatun Formation of northern Panama contains a highly diverse and well sampled neritic fossil assemblage that was located in the Central American Seaway that connected the Pacific and Atlantic (Caribbean) oceans ~10 million years ago. Based on field discoveries and analysis of existing collections, the sharks from this unit consist of at leat 16 taxa, including four species that are extinct today. The remaining portion indicates relatively long-lived species. Based on the known habitat preferences for modern selachian, the Gatun sharks were primarily adapted to shallow waters within the neritic zone. Also, in comparison with modern species, the Gatun shark fauna has mixed Pacific-Atlantic biogeographic affinities. Comparisons of Gatun dental measurements with other formations suggest that many of the species have an abundance of small individuals. One of this small-size species is the extinct *Carcharocles megalodon*, paradoxically the biggest shark that ever lived. Here, the tooth sizes from the Gatun Formation were compared with isolated specimens and tooth sets from different aged, but analogous localities. In addition, the total lengths of the individuals were calculated. This comparisons and estimates suggest that the small size of Gatun's *C. megalodon* is not related to timing (chronoclines), or to the tooth position within a variant jaw, and that the individuals from Gatun were mostly juveniles and neonates.

Pimiento's dissertation marks an important change in empirical culture. Alongside a traditional research project (concerning the biodiversity of late Miocene sharks and the possibility of a *Carcharocles megalodon* nursery in the Central American Seaway), Pimiento presents *Fossil Sharks from Panama*: an interactive bilingual website dedicated to the geological, biological and ecological education of ancient Panamanian sharks to an elementary and middle school audience. This website is founded on the studied principle that, in modern times, "popular media, in the form of radio, television and the Internet, make science information gradually more available to people across venues for science learning." Pimiento is wise to note that linear education (i.e., in the form of textbooks or lecture) is more static and less attractive than dynamic, interactive education (i.e., field experiences, interactive games, and social media).

The modern global age exists with an unprecedented degree of intellectual freedom and satisfaction: media can be quick to placate most queries or curiosities (be they intensive or spur-of-the-moment), and that is a precious gift completely unique to the most recent of generations. However, the overwhelming degree of media exposure has simultaneously selected for a shrinking attention span: national news, video games, commercial products, intellectual discoveries, celebrity gossip, artistic exhibitions—one may switch between any of these subjects within a single minute online or in front of television. With such a revolution on intellectual dissemination, science and education face a crucial crossroads: how can such institutions compete for attention within a modern ocean of media distraction, and yet still *impress* themselves upon their students?

Pimiento is looking forward. For information on Miocene *Carcharhinus leucas,* please take her modern map:

[http://www.stri.si.edu/english/kids/sharks/dientes.html].

2011

Catálogo Ilustrado dos Tubarões e Raias dos Açores –
Sharks and Rays from the Azores an Illustrated Catalogue
IAC-Instituto Açoriano de Cultura,
ISBN 978-989-8225-24-5
BARREIROS, J.P. & GADIG, O.B.F.

DESCRIPTION

Another typical "requiem" shark, although particularly robust and massif with a short snout. As in the previous species no longitudinal dermal crest between dorsal fins. Fins with pointed or slightly rounded apex. Upper teeth triangular and serrated. Upper body and flanks grey and belly white with no white spots or blotches on fins.

FOOD

Fish, other sharks, rays, shrimp, crabs, squid, shellfish, sea urchins, carrion, sea turtles and occasionally, garbage and all sort of human dump. In the Ganges River it regularly feeds on human corpses and, in places, is responsible for several fatal attacks on bathers, surfers and some divers.

REPRODUCTION

Placental viviparous producing up to 13 newborns per litter.

DISTRIBUTION

Common in warm seas worldwide mainly between 42°N, 39°S; 117°W, 155°E. Occurs in coastal waters, rivers and estuaries. In some big rivers (Amazon, Zaire, Zambezi, Mississippi, Ganges, Indus, etc.) it is often found several miles upwards. Some populations even live in fresh water lakes, the best known being that from Lake Nicaragua (where it was described as a different species – *C. nicaraguensis*).

Did you catch that? Arguably the most powerful clause yet written by an empirical master: "In the Ganges River it regularly feeds on human corpses". Disturbing, yes? We have talked before about *Carcharhinus leucas* as the ultimate example of heterotrophy: life begat from death. However, the introduction of a new human element—the corpse— accentuates this formidable dynamic; in the English language, 'corpse' is a particularly unnerving term, whose simple usage encourages an eruption of thought...Corpse is the soulless shell—the departed vessel, whose occupant dissolves into the ultimate of death. Corpse is the final bridge between the physical world and the unknown, the impossible, the spiritual; it is the mark of death itself, and the most powerful practical proof of its imminence.

Death is the maker of convictions. Death is the core of worldviews and spiritualties. Death has a hand in politics, in satisfaction, in personal development: it is nature's most powerful urgency. Death is most strongly evoked with the emptiness of a corpse, and it is this corpse—in all of its gravitas and implication—that lies within the jaws of 'Life Begat from Death': the jaws of *C. leucas*. Is this not equally marvelous and horrific? That for sustenance the Bull Shark elects a human corpse? What other animal can have this option—the power to so completely entrap our souls?

Such imagery is somehow progressed from that of the classic Shark Attack. *Carcharhinus leucas* is here not the cause of death: it is the devourer of its mark. Yet in both cases, the Bull Shark acts as the physical destiny; for an extremely rare portion of humanity, the Bull Shark's jaws are the final stop before disincorporation. Who among us will share in that destiny? Who could imagine the end (or the beginning) at such a bloody threshold? *Carcharhinus leucas* elicits these questions, and it is one of the few beings that can. How truly terrific...that a Bull Shark could end my life and take its corpse...

2011

The Presence, Source and Use of Fossil Shark Teeth from
Ohio Archaeological Sites
Ohio Archaeologist, 61 (4): 26-46
COLVIN, G.H.

Fossil shark teeth are the most commonly collected vertebrate
fossil in the world (Hubbell 1996). The abundance of shark teeth as fossils
is due to a shark's rapid and continual loss and replacement of teeth, the
relative ease in which the teeth fossilize, and the presence of sharks in our
oceans for the last 400 million years. Modern studies of living sharks
indicate that a shark may shed and replace their 50 to 100 biting teeth as
often as every seven days (Purdy 1998). Once lost, the teeth fall to the sea
floor where they are buried and often fossilized. Whether for curiosity or as
a tool, early Americans have collected fossil shark teeth for at least 10,000
years (Royal and Clark 1960). Although rare, fossil shark teeth have been
found at archaeological sites in Ohio. Based on the relatively young age of
these fossil shark teeth (Cenozoic Era) and the much older rock formations
in Ohio (Paleozoic Era), we can, in most cases, be confident that the source
of the fossil shark teeth found at Ohio archaeological sites is not Ohio. This
means that these shark teeth must have been brought into Ohio by
prehistoric people.

Fossil shark teeth can be found in abundance in rivers,
unconsolidated marine deposits, and beaches on the east coast of the United
States and the Gulf of Mexico. Squier and Davis (1848) were the first to
formally report and question the source of fossil shark teeth from
archaeological sites within the Mississippi Valley.

Many Americans are familiar with Polynesian shark culture, wherein sharks were worshipped as gods or legends; however, it is rather ironic that many are simultaneously unaware of Hopewellian shark culture—a comparably complex tradition that had originated from the valleys and coasts of the Eastern United States. Archaeological investigations into the Ohio Hopewell culture have yielded a number of fossilized shark teeth, each dating to a geological period much younger than that of the surrounding Ohio landscape. It is thus most likely that the Hopewell culture traded for these teeth with other tribes situated near fossil deposits on the North American Gulf and Atlantic coasts.

Modern shark teeth make for effective cutting and carving tools, and could have been weaponized as spearheads or arrowheads; they too emerge from Hopewell archaeological sites, and were likely acquired from trade with ancient shark hunting cultures. Fossil teeth were less effective as practical tools but more revered as spiritual objects and adornments. Indeed, many fossil teeth (especially those of the Megalodon) have been discovered with perforations drilled near the base (usually at the nutrient pore: a small groove or hole near the center of the root). These perforations were most likely threaded to form a necklace—an object of spiritual authority. It is unlikely that such necklaces were worn for ordinary purposes, as the tooth itself may mimic Hopewellian shamanic transformation: the shark tooth's physical juxtapositions of light and dark (modern and fossil) and dull and shiny (root and crown) were appropriate symbols of this tradition.

C. leucas teeth are uncommon at Hopewell sites, and probably originated from Florida. Fossil teeth have been found with drilled perforations, and thus may have served shamanic purposes (two undrilled teeth have also been found near Mound City). Modern variants of the lower teeth are naturally shaped for awl and drills, and thus could have been used.

2011

La Pesqueria de Tiburones Oceanicos-Costeros en los Litorales de Colima, Jalisco y Michoacan
Revista de Biología Tropical, 59 (2): 655-667
CRUZ, A. & SORIANO, S.R. & SANTANA, H. & RAMIREZ, C.E. & VALDEZ, J.J.

Shark fishery is one of the most important activites in the Mexican Pacific coast, nevertheless, there is few data available about the specific captures done by the fleet along the coast. This study describes fishery biology aspects of the shark species catched by the semi-industrial long-line fleet of Manzanillo. Monthly samplings were made on board of these vessels during an annual period from April 2006 to April 2007. Captured species composition (n=1 962 organisms) was represented by nine species. The one that sustains this fishery was *Carcharhinus falciformis* (88.12%), followed by *Prionace glauca* (8.21%). Low frequency species were represented by *Sphyrna zygaena* (1.78%), *Alopias pelagicus* (0.82%), *Carcharhinus longimanus* (0.45%). Furthermore, rare species were *Alopias superciliosus* (0.35%), *Carcharhinus leucas* (0.1%), *Carcharhinus limbatus* (0.1%) and *Isurus oxyrinchus* (0.05%). Fishery activity affected principally (60-92.70%) young males of *C. falciformis*, *S. zygaena*, *C. longimanus* and *I. oxyrinchus*; adult males (56-75%) of *A. pelagicus*, *A. superciliosus*, and *C. limbatus*; for *P. glauca* there were primarily female adults. For all the species found, females showed the bigger sizes when compared to males (with the exception of *S. zygaena*, that showed sexual dimorphism).

Palabras clave: long-line fleet, Manzanillo, fishery, pregnant females, embryo.

Only two Bull Sharks were captured in this study (a 110 cm juvenile and a 271 cm adult) off the coast of Michoacan. Though incredibly rare, *Carcharhinus leucas* is pound-for-pound the third-most valuable species in the Mexican shark fishery; only *C. falciformis* (Silky Shark) and *Sphyrna zygaena* (Smooth Hammerhead) command a higher price. Commercially, these sharks are harvested for their meat and fins; however, local fisheries additionally extract their liver oil through traditional practices, and furthermore utilize their teeth and jaws for local handicrafts.

Approximately 100 shark species can be found in Mexican waters, but of these, only 40 are incorporated into the commercial fisheries. *Carcharhinus falciformis* is by far the most heavily utilized, and serves as the bedrock of the Mexican catch (as it is simultaneously the most abundant species and most valuable pound-for-pound). Silky Sharks are captured throughout the entire year, and can be commonly found in both oceanic and coastal environments. Of concern is the species' lower fertility rate (as compared to other Carcharhinids), which increases its vulnerability to overexploitation.

What is the principal difference between the Bull Shark and the Silky Shark (or the Sandbar Shark for that matter), that one should be so rare in North American fisheries while the other abounds? It is by now a rather tired question, as we are acutely aware of the Bull Shark's rarity in the American market; to observe such a comparable rarity in the Mexican fishery is to simply make the quandary even more intriguing. As *Carcharhinus leucas* is ecologically common (and furthermore has extensively documented American nurseries), I feel that its rarity is most likely related to either fisheries method or feeding behavior. Could it be that the longlines are simply 'too pelagic' for this species—a shark that has a strong (and demonstrable) preference for patrolling the coastal shallows?

2011

Los Tiburones Carcharhiniformes (Chondrichthyes, Galeomorphii) del Plioceno Inferior de la Formación Arenas de Huelva, Suroeste de la Cuenca del Guadalquivir, España

Revista Mexicana de Ciencias Geológicas, 28 (3): 474-492

GARCÍA, E.X.M. & BALBINO, A.C. & ANTUNES, M.T. & RUIZ, F. & CIVIS, J. & ABAD, M. & TOSCANO, A.

Carcharhiniform shark teeth from eight localities of the lower Pliocene Arenas de Huelva Formation were studied in the southwestern part of the Guadalquivir basin. Thirty-five samples were collected. Three hundred kilograms of sediments were levigated and 48 teeth were found. The genera *Megascyliorhinus*, *Premontreia*, *Mustelus*, *Paragaleus*, *Carcharhinus* and *Galeocerdo* were recognized. *Carcharhinus* diversity and abundance prevail followed by *Mustelus* and *Premontreia*, whereas the other taxons are scarce. The selachians assemblage points out to littoral-neritic, subtropical to temperate environments. Carcharhiniform genera that have been found in the lower Pliocene of the Guadalquivir Basin also were found in the upper Miocene of the Alvalade Basin in Portugal.

Key words: Chondrichthyes, Carcharhiniformes, Lower Pliocene, Huelva, Spain

The following table describes the stratigraphy of the Guadalquivir basin, which includes a fossil assemblage of: Foraminifera (*F*), scallops and clams (*S*), Oysters (*O*), undifferentiated marine fauna (*U*), gastropods (*G*), barnacles (*B*), *Clypeaster* sea biscuits (*C*), chondrichthyan and osteichthyan teeth (*T*), otoliths (*L*) and vertebrate bones (*V*).

AGE	FORMATION	*F*	*S*	*O*	*U*	*G*	*B*	*C*	*T*	*L*	*V*
Upper Pliocene	Fm. Arenas de Bonares		S	O			B				
Lower Pliocene	Fm. Arenas de Huelva		S	O	U	G			T	L	V
Miocene (Messin.)	Fm. Arcillas de Gibraleón	F	S			G			T		
Miocene (Torton.)	Fm. Calcarenita de Niebla		S	O			B	C	T		

Carcharhinus leucas was specifically found in the localities of Casa del Pino (37°20'0.40"N; 6°40'33.63"W) and Ambulatorio (37°19'38.66"N; 6°40'37.39"W) near the town of Bonares (in Spain's Huelva province). The Arenas de Huelva formation in which the teeth were discovered is largely composed of fine sands and mud, and yields a rich diversity of subtropical and warm-temperate paleofauna dating from the lower Pliocene. Though characteristically subtropical sharks such as *C. perezi* compliment this formation, García et al. are quick to note the relative absence of *Galeocerdo*—a famously subtropical genus (even in the Mediterranean). Only one *Galeocerdo* tooth (*G. aduncus*, Casa del Pino, Bonares) was found in this study. Could *Galeocerdo* have been the first Carcharhinid to gradually emigrate from the ancient Mediterranean? If so, it appears that *C. leucas* and *C. perezi* followed suit soon after—but what was the order of their departure? Could the Bull have been last?

2011

Shark Attacks on Irrawaddy Dolphin in
Chilika Lagoon, India
Journal of the Marine Biological Association of India
53 (1): 27-34
KHAN, M. & PANDA, S. & PATTNAIK, A.K. & GURU, B.C. &
KAR, C. & SUBUDHI, M. & SAMAL, R.

Irrawaddy dolphin *Orcaella brevirostris* Gray, 1866 inhabit Chilika lagoon, Odisha on the east coast of India. The narrow outer channel connecting the main lagoon and the Bay of Bengal harbor approximately 60% of the Irrawaddy dolphin population. The bull shark *Carcharhinus leucas* Müller and Henle, 1839 travels from the sea 5-10 km inward to the outer channel during late gestation period. Mortality of Irrawaddy dolphins related to bull shark attack ranged from 1 to 3% of the population. An account of the shark attack on Irrawaddy dolphin was first recorded on 24th April, 2003 from the Chilika lagoon. Salinity and depth parameters of the outer channel are conducive for both the shark and the dolphin show the overlapping of the habitat. This study indicates that habitat or prey overlap may be responsible for the fatal bull shark attacks on Irrawaddy dolphin in Chilika lagoon.

Keywords: Irrawaddy dolphin, bull shark, shark attack

The Irrawaddy Dolphin is small, spurtive and unbearably cute. It has been nicknamed, "the smiling face of the Mekong", and is known to help fishermen by driving schools of fish into their nets (in exchange for a scrumptious reward). Our mammalian tendency is to nurture beings of such a pudgy, soft and smiling nature; even though we may not truly *understand* (and by extension, sincerely love) the Irrawaddy Dolphin, we have a compelling urge to care for this adorable species and its wellbeing. As a mirror to this 'cute bias', we abhor the account of Khan et al.—an exquisitely detailed analysis of the Bull Shark attack in all of its gruesome, awesome and upsetting power.

Pregnant *Carcharhinus leucas* enter the Dolphin's habitat in Chilika Lagoon around December or January, and remain for roughly 6 or 7 months (departing in June). Throughout most of this period of cohabitation, Bull Sharks and Irrawaddy Dolphins feed on a very similar assemblage of prey items (70% of prey species are common to both predators). Khan et al suggest that these two species compete within a similar predatory niche, and further claim that this competition "may be the important reason for such fatal shark attacks on Irrawaddy dolphins." Whatever the reason, the attacks are devastating.

C. leucas will launch itself into the dolphin's soft underbelly and bite between the thoracic and uro-genital regions. One such bite can rip off up to 10kg of, "skin, blubber, muscles, rib bones...and portion of intestine". The impact and devastation of this bite allegedly causes, "rupture and serration in arteries and veins of abdomen, leading to massive haemorrage before death." In two of Khan's dolphins, Bull Shark teeth have been found lodged into the ribs. Instinctively, such brutality plays our empathy; we find ourselves subject to a visceral realm—a place where the Bull Shark exists as little more than a monster. Thus, we too are victims of the bite.

2012

Field Guide to Sharks of the Southeast Asian Region
SEAFDEC, 2012
AHMAD, A. & LIM, A.P.K.

The sharks landing comprise only a small percentage of the total marine fishes in the Southeast Asian Region. However they provide significant incomes for traditional fishers. They have been a cheap source of protein for poor people in remote areas as well as coastal communities.

Indonesia, Malaysia and Thailand are three major countries in this region recorded high catch of sharks. For centuries, fishers in these countries have conducted fishing for this resource sustainably and some still do. However, in the recent decades, the advent of modern fishing vessel and its technology which could access distant fishing ground have cause an increased in effort and yield of catches, as well as an expansion of the fishing areas. As a result of overexploitation, several species and some stocks are said to be endangered in several areas.

Sharks and rays become one of the majort international fisheries issues since the late 90's when several sharks species was proposed to be listed in CITES Appendixes. Many NGOs that are very concerned with the environment as well as animals are actively campaigning for more effective measures to be taken to conserve, manage and protect them from being exploited by unfriendly fishing gears.

In term of taxonomy, Class Chondrichthyes include sharks, rays, skates, chmareas and elephant fish. These fishes differ from the Osteichthyes or bony fishes as they possess a cartilaginous skeleton instead of a bony skeleton.

Carcharhinus leucas is the fourth most commonly harvested shark in Cambodia, and has an extensive presence in the Southeast Asia region as defined by SEAFDEC (Southeast Asian Fisheries Development Center). Though absent in Vietnam, Bull Sharks can be found in Brunei Darussalam, Indonesia, Malaysia, Myanmar, Thailand and the Philippines. Ahmad & Lim's field guide offers little new biological information on the species, but compensates with an unprecedentedly rich diversity of regional names:

English	Bull shark ~ River whaler ~ Freshwater whaler
Malay	Yu garang ~ Jerung sapi
Thai	Chalarm Hua-bart
Indonesian	Cucut bekeman ~ Hiu buas ~ Hiu bujit
Japanese	Oomejiro zame
Cambodian	Ka Mab
Myanmar	Nga-mann

This may be our first publication in which the authors acknowledge a global total of 500 shark species (the current number, as of September 1st, 2016, is 512 sharks). Ahmad & Lim wisely predict that this number will be, "expected to increase in the future." Advances in genetics and deepwater technology have greatly enriched our present understanding of global shark biodiversity and will most likely champion future taxonomic expansion. As of yet, we have only just begun to probe the world's deepwater canyons, shelves and abyssal plains; this is particularly true for Southeast Asian seas, as Ahmad & Lim confirm that, "deep water species are mostly unknown due to limited research activity." So far, a total of 174 shark species have been found in Southeast Asia.

2012

Squalicorax Chips a Tooth: A Consequence of Feeding-Related Behavior from the Lowermost Navesink Formation (Late Cretaceous: Campanian-Maastrichtian) of Monmouth County, New Jersey, USA
Geosciences, 2 (2): 109-129

BECKER, M.A. & CHAMBERLAIN, J.A.

Chipped and broken functional teeth are common in modern sharks with serrated tooth shape. Tooth damage consists of splintering, cracking, and flaking near the cusp apex where the enameloid is broken and exposes the osteodentine and orthodentine. Such damage is generally viewed as the result of forces applied during feeding as the cusp apex impacts the sekeletal anatomy of prey. Damage seen in serrated functional teeth from sharks *Squalicorax kaupi* [1] and *Squalicorax pristodontus* [1] from the late Cretaceous lowermost Navesink Formation of New Jersey resembles that occurring in modern sharks and suggests similar feeding behavior. Tumbling experiments using serrated modern and fossil functional shark teeth, including those of *Squalicorax*, show that teeth are polished, not cracked or broken, by post-mortem abrasion in lowermost Navesink sediment. This provides further evidence that chipped and broken *Squalicorax* teeth are feeding-related and not taphonomic in origin. Evolution of rapid tooth replacement in large sharks such as *Squalicorax* ensured maximum functionality after feeding-related tooth damage occurred. Serrated teeth and rapid tooth replacement in the large sharks of the Mesozoic and Cenozoic afforded them competitive advantages that helped them to achieve their place as apex predators in today's ocean.

Becker & Chamberlain's selection of *Carcharodon carcharias* (White Shark), *Galeocerdo cuvier* (Tiger Shark) and *Carcharhinus leucas* as the best modern physical examples of *Squalicorax* is an unintentionally profound comment on evolutionary history and its inexorable relationship with humanity's common understanding of sharks. These three modern species had *evolved* serrated teeth in the pursuit of larger prey; with such a unique adaptation, Ancient Bull Sharks (and their serrated comrades) could successfully exploit new niches involving larger prey items such as, "marine mammals, turtles, birds, large osteichthyans and other chondrichthyans." The success of this adaptation secured the modern White, Tiger and Bull sharks as *uniquely* apex predators, and reinforced their development into powerful, massive and opportunistic hunters. By the time *Homo sapiens* began to construct coastal settlements, these three sharks were more than ready to capitalize on their natural talents: thus began our common understanding of "Shark".

It is no coincidence that the White Shark, Tiger Shark and Bull Shark are by far the most popular (and most feared) selachians. We, like all species, have evolved to address threats and mitigate risk, and when faced with such opportunistic hunters, we form a complex pattern of survival. We began this pattern with simple acknowledgment: "don't go into the water; there is a shark, and it will eat you." But then, as we developed more constructs—as we incorporated religions, philosophies, stories and art into our lives—we transformed the pattern. "Shark" became a god. "Shark" became a demon. "Shark" became a specific form—a streamlined, toothy cross between a White and a Bull. Why is it that we hardly consider the other forms—the nurse, bullhead, angel, lantern, frilled and other sharks—as selachians equally worthy of our cultural airtime? It is simply because they are harmless: they didn't evolve the teeth that touched humanity.

2012

A Comparison of Spatial and Movement Patterns between
Sympatric Predators: Bull Sharks (Carcharhinus leucas)
and Atlantic Tarpon (Megalops atlanticus)
PLoS ONE, 7 (9): e45958
HAMMERSCHLAG, N. & LUO, J. & IRSCHICK, D.J. &
AULT, J.S.

Predators can impact ecosystems through trophic cascades such that differential patterns in habitat use can lead to spatiotemporal variation in top down forcing on community dynamics. Thus, improved understanding of predator movements is important for evaluating the potential ecosystem effects of their declines. We satellite-tagged an apex predator (bull sharks, *Carcharhinus leucas*) and a sympatric mesopredator (Atlantic tarpon, *Megalops atlanticus*) in southern Florida waters to describe their habitat use, abundance and movement patterns. We asked four questions: (1) How do the seasonal abundance patterns of bull sharks and tarpon compare? (2) How do the movement patterns of bull sharks and tarpon compare, and what proportion of time do their respective primary ranges overlap? (3) Do tarpon movement patterns (e.g., straight versus convoluted paths) and/or their rates of movement (ROM) differ in areas of low versus high bull shark abundance? And (4) Can any general conclusions be reached concerning whether tarpon may mitigate risk of predation by sharks when they are in areas of high bull shark abundance?

Bull shark abundance was high year-round, but peaked in winter; while tarpon abundance and fishery catches were highest in late spring. However, presence of the largest sharks (>230 cm) coincided with peak tarpon abundance.

Florida is rather famous for a unique fishing experience: the climatic and bloody thievery of a Bull Shark devouring a "Silver King" on the hook. This awesome display of brutal predatory might is merely a snapshot of a complex dynamic: the trophic ballet of *Carcharhinus leucas* and *Megalops atlanticus*. Bull Sharks and Tarpon occupy a very similar predatory niche on the Florida coast (wherein each preys upon mullet, menhaden and ladyfish with equal voracity). However, Tarpon find themselves at risk of Bull Shark attack when they forage within this niche: as such, the "Silver Kings" appear to undertake extraordinary measures in order to mitigate this risk of *Carcharhinus leucas* predation.

Bull Sharks exhibit high-site fidelity in Florida's shallow inshore waters, and, "are commonly observed preying upon tarpon at popular fishing locations in the Florida Keys, southern Florida and Gulf of Mexico". Of particular interest is a 670 km^2 "core area" for Bull Sharks located in the northwestern part of Florida Bay. Hammerschlag et al. propose that this hotspot is likely a productive feeding ground, and may furthermore be the epicenter of a behaviorally mediated indirect interaction (BMII) between Bull Sharks and Tarpon. Specifically, the authors propose that, "higher shark abundance in the northwestern area of Florida Bay is largely driven by relatively high teleost abundance (preferred prey) there, which in turn, indirectly causes tarpon to reduce their use of this productive area when foraging to minimize their risk of potential mortality by sharks"

Avoidance of this "core area" is but one of the Tarpon's sacrificial stratagems (for while the fish is safe, it pays an opportunity cost in the form of forgoing the foraging ground). Another such compromise is the Tarpon's ascent into osmotically challenging habitats (such as rivers). Though adult Bull Sharks could indeed pursue, they prefer to remain in osmotic balance with coastal waters (so as to not incur osmoregulatory costs).

2012

*The Conservation Status of North American, Central
American, and Caribbean Chondrichthyans*
IUCN Species Survival Commission Shark Specialist
Group, Vancouver, Canada
KYNE, P.M & CARLSON, J.K. & EBERT, D.A. & FORDHAM,
S.V. & BIZZARRO, J.J. & GRAHAM, R.T. & KULKA, D.W. &
TEWES, E.E. & HARRISON, L.R. & DULVY, N.K.

Bull Shark

Carcharhinus leucas (Valenciennes, *in*
Müller & Henle, 1839)

Red List Assessment: Global: Near Threatened
(Simpfendorfer, C.A. & Burgess, G.H.
2000).

Regional Occurrence: Northwest Atlantic; Western Central
Atlantic; Eastern Central Pacific.

Rationale: The Bull Shark (*Carcharhinus leucas*) is
a common tropical and subtropical
species that occurs in marine, estuarine
and fresh waters. It regularly penetrates
long distances up large rivers. It is caught
in fisheries throughout its range, but it is
rarely a target species. Its occurrence in
estuarine and freshwater areas makes it
more vulnerable to human impacts and
habitat modification.

The IUCN lists *Carcharhinus leucas* as Near Threatened (NT), as it is a species that does not qualify for (or is likely to qualify for) a threatened category now, but is close to qualifying for (or is likely to qualify for) a threatened category in the near future. In some ways this a relief, as the threatened categories correspond to an extinction risk:

VU	Vulnerable	High risk of extinction in the wild
EN	Endangered	Very high risk of extinction in the wild
CR	Critically Endangered	Extremely high risk of extinction in the wild

However, concern for the Bull Shark's wellbeing as a species is mounting as populations are beginning to experience the pressures that have pushed many a fellow *Carcharhinus* into IUCN Vulnerable (VU) status or worse. In the Caribbean Sea, overexploitation has driven *Carcharhinus leucas* away from its nearshore coastal habitats and towards outlying barrier reefs; this distributional shift has significantly altered the local ecological regime, as Blacktip and Caribbean Sharpnose Sharks have, "populated the vacated coastal niches and now represent the majority of shark captures in nearshore waters." In the Gulf of California, *Carcharhinus leucas* is named as one of the many large sharks experiencing major population declines (the geographical extent of which is not fully understood, as the species' American Pacific range has yet to be elucidated).

Beyond the Americas, the Bull faces dams, protective meshes and other destructive alterations thanks to its perilous proximity to ever-encroaching human habitat. Though global initiatives such as the IUCN, FAO and CITES have greatly enriched our conservation mindset, we still must consider their power of protective action in the face of growing human development: is it not our *inflation* that most wounds the Bull Shark?

2012

Cyanobacterial Neurotoxin Beta-N-Methylamino-L-Alanine (BMAA) in Shark Fins
Marine Drugs, 10 (2): 509-520
MONDO, K. & HAMMERSCHLAG, N. & BASILE, M. & PABLO, J. & BANACK, S.A. & MASH, D.C.

Sharks are among the most threatened groups of marine species. Populations are declining globally to support the growing demand for shark fin soup. Sharks are known to bioaccumulate toxins that may pose health risks to consumers of shark products. The feeding habits of sharks are varied, including fish, mammals, crustaceans and plankton. The cyanobacterial neurotoxin β-N-methylamino-L-alanine (BMAA) has been detected in species of free-living marine cyanobacteria and may bioaccumulate in the marine food web. In this study, we sampled fin clips from seven different species of sharks in South Florida to survey the occurrence of BMAA using HPLC-FD and Triple Quadrupole LC/MS/MS methods. BMAA was detected in the fins of all species examined with concentrations ranging from 144 to 1836 ng/mg wet weight. Since BMAA has been linked to neurodegenerative diseases, these results may have important relevance to human health. We suggest that consumption of shark fins may increase the risk for human exposure to the cyanobacterial neurotoxin BMAA.

Keywords: β-N-methylamino-L-alanine; neurotoxin; neurodegenerative disease; cyanobacteria; elasmobranch; conservation

How magnificent it is, that one so small could be Creator and Destroyer. Cyanobacteria are famous for their photosynthetic oxygenation of the Earth; it is they who have dictated our physical dependence on Air, and it is they who now threaten our cognitive existence with the debilitating neurotoxin, BMAA (which is linked to a number of neurodegenerative diseases, such as Alzheimer's and ALS). Though only certain free-living species produce BMAA, they can be easily ingested; filter feeders, zooplankton and even free-swimming nekton (unfortunate enough to pass through cyanotoxic waters) are all susceptible to the poison. As such, BMAA is able to infiltrate every trophic level.

We usually envision the top of the food chain as the most powerful trophic position, but perhaps it is the most vulnerable. *Carcharhinus leucas* and other apex predators are at the end of a long transaction of energy, atoms and everything in between: by consuming even a single tarpon, the Bull Shark consumes everything the tarpon has eaten (including mullet, menhaden, crabs, shrimp and any planktonic organisms that they too have ingested within their own predatory niches). BMMA presence at any one of these levels is transferrable; poisoned zooplankton can be passed upward to menhaden, which can then transfer their toxin to tarpon (and, by final extension, to Bull Sharks).

As all trophic roads lead to the apex, *Carcharhinus leucas* accumulates all poisons; BMMA, much like mercury, concentrates at higher levels (rendering large predatory sharks like the Bull as some of the riskiest catches to consume). Indeed, Mondo et al. report BMAA presence in the fins of two *Carcharhinus leucas* from Florida Bay (in the midst of a cyanobacterial bloom, a 163 cm shark bore a BMAA concentration of 232 ng/mg, while a 183 cm individual bore a higher concentration of 264 ng/mg). If ingested, such Bulls could bridge a demented fate.

2012

A DNA Sequence Based Approach to the Identification of Shark and Ray Species and its Implications for Global Elasmobranch Diversity and Parasitology
Bulletin of the AMNH, 367: 262 pp., 102 figures, 5 tables
NAYLOR, G.J.P. & CAIRA, J.N. & JENSEN, K. & ROSANA, K.A.M. & WHITE, W.T. & LAST, P.R.

In an effort to provide a framework for the accurate identification of elasmobranchs, driven in large part by the needs of parasitological studies, a comprehensive survey of DNA sequences derived from the mitochondrial NADH2 gene was conducted for elasmobranchs collected from around the world. Analysis was based on sequences derived from 4283 specimens representing an estimated 574 (of ~1221) species (305 sharks, 269 batoids), each represented by 1 to 176 specimens, in 157 (of 193 described) elasmobranch general in 56 (of 57 described) families of elasmobranchs (only Hypnidae was not represented). A total of 1921 (44.9%) of the samples were represented by vouchers and/or images available in an online host specimen database (http://elasmobranchs.tapewormdb.uconn.edu).

A representative sequence for each of the 574 species identified in this survey, as well as an additional 11 sequences for problematic complexes, has been deposited in GenBank. Neighbor-joining analysis of the data revealed a substantial amount of previously undocumented genetic diversity in elasmobranchs, suggesting 79 potentially new taxa (38 sharks, 41 batoids). Within-species p-distance variation in NADH2-percent sequence divergence ranged from 0 to 2.12 with a mean of 0.27; within-genus p-distance variation ranged from 0.03 to 27.01, with a mean of 10.16.

Mark this point: Naylor et al., 2012—the death of *Carcharhinus leucas* as we knew it. Beforehand, we have accumulated many identities into one totality: the Nicaragua Shark, Zambezi Shark, Cub Shark and others were merely faces of *Carcharhinus leucas*—the single, final, truthful Bull Shark. We have for decades conducted research under this tidy pretense—a truly sensible determination. But signs existed: regional discrepancies in vertebral counts, uncertainty in regards to the genetic distinctiveness between populations— these were classic warnings that the Bull Shark as we knew it was something not fully resolved—something *complex*. Until recently, we have lacked the tools to make such determinations; but now, with genetics...

Our entire understanding of the species' identity is shattered: *Carcharhinus leucas* may not be one shark, but instead, a complex of *three* distinct species. Naylor et al. determined three potential genetic subclusters from 24 specimens originally identified as *Carcharhinus leucas*. The first subcluster – *Carcharhinus leucas* – includes specimens from the Western Atlantic, Belize, Sengal and Sierra Leone (as Müller & Henle's original type specimen of *Carcharhinus leucas* came from the Antilles, Naylor et al. suggest that the original binomial most appropriately applies to this Atlantic group). The second subcluster – *Carcharhinus* cf. *leucas* 1 – applies to seven specimens from Borneo, while the third subcluster – *Carcharhinus* cf. *leucas* 2 – applies to three specimens from South Africa. Although notable variations in haplotype exist within each of these three subclusters, there is, "no overlap in haplotypes among the three potential species of bull sharks." Furthermore, Naylor et al. determine that (much in favor of speciation), the "haplotypes of all three bullshark species are allopatrically distributed." While further investigations must be conducted, it seems that the Bull Shark Triumvirate is immanent.

2012

Natural or Artifical? Habitat-Use by the Bull Shark,
Carcharhinus leucas
PLoS ONE, 7 (11): e49796
WERRY, J.M. & LEE, S.Y. & LEMCKERT, C.J. & OTWAY,
N.M.

Despite accelerated global population declines due to targeted and illegal fishing pressure for many top-level shark species, the impacts of coastal habitat modification have been largely overlooked. We present the first direct comparison of the use of natural versus artificial habitats for the bull shark, *Carcharhinus leucas*, an IUCN 'Near-threatened' species – one of the few truly euryhaline sharks that utilizes natural rivers and estuaries as nursery grounds before migrating offshore as adults. Understanding the value of alternate artificial coastal habitats to the lifecycle of the bull shark is crucial for determining the impact of coastal development on this threatened but potentially dangerous species.

We used longline surveys and long-term passive acoustic tracking of neonate and juvenile bull sharks to determine the ontogenetic value of natural and artificial habitats to bull sharks associated with the Nerang River and adjoining canals on the Gold Coast, Australia. Long-term movements of tagged sharks suggested a preference for the natural river over artificial habitat (canals). Neonates and juveniles spent the majority of their time in the upper tidal reaches of the Nerang River and undertook excursions into adjoining canals. Larger bull sharks ranged further and frequented the canals closer to the river mouth.

Gold Coast is Australia's fastest growing city and a beacon for tourism; as a sprawling seaside paradise, the city features an extensive network of canals—the Gold Coast System (GCS)—intended to accommodate the increasing spread of urbanization and demand for gorgeous waterfront property. As the GCS was originally carved from the Nerang River—an important tidal estuary—its many present passages link directly to a rich community of species benefitting from the Nerang's estuarine bounty and protective mangroves. All such denizens have access to the GCS canals—even young and developing Bull Sharks.

The natural Nerang estuary system is preferred by all life history stages. During the austral spring and summer, pregnant females travel upriver into the Nerang's reduced salinities for parturition. Neonates exhibit high site fidelity within a rather restricted area, and largely benefit from the natural protection of mangroves and their structural refuge. Neonates thus rarely enter the GCS canals; but when they transition into the juvenile stage, they become emboldened. Juvenile Bull Sharks outgrow their mangrove shelters to embrace varying salinities, explore wider ranges and engage larger prey items. Unlike the harmless neonates, potentially dangerous juveniles frequently enter the artificial canal habitat (perhaps as a means to avoid larger, oceanic predators—including cannibalistic adult Bull Sharks—via the structural complexity and spatial expanse of the canals).

Human intrusion into Bull Shark habitat is thus doubly negative. Werry et al. affirm that while Bulls are obviously pained from such critical development (for the canals not only directly remove habitat, but also introduce new destructive pressures such as fishing, anti-fouling paints and urban runoff) humans too suffer from an increased, man-made risk of Bull Shark attack. No stage of Bull Shark would prefer the canals to the natural Nerange; however, it seems this hardy species is keen to make-do.

2013

Visual Identification of Fins from Common Elasmobranchs in the Northwest Atlantic Ocean
NOAA Technical Memorandum NMFS-SEFSC-643: 51pp
ABERCROMBIE, D.L. & CHAPMAN, D.D. & GULAK, S.J.B. & CARLSON, J.K.

Shark fins are a highly prized commodity in many Asian cultures (Fong and Anderson, 2000; Clarke et al., 2007). Sharks sourced for fins come from directed fisheries, from bycatch in multi-species fisheries, and illegal, unreported and unregulated (IUU) fisheries (Lack and Sant, 2008). Despite the fact that expanding global demand for shark products (especially fins) has resulted in substantially increased exploitation of sharks worldwide, utilization and trade in shark parts remain poorly assessed in terms of its species composition and magnitude. The limited data available are often difficult to interpret and are often not accurate enough to indicate which species and regions are most affected by the trade (Rose, 1996; Clarke, 2002; Clarke, 2004). In response to the international trade in fins and other products, the status of several elasmobranch species, the white (*Carcharodon carcharias*), basking (*Cetorhinus maximus*) and whale shark (*Rhincodon typus*) have been listed in Appendix II and 4 species of sawfish (Family Pristidae) in Appendix I of the Convention on International Trade in Endangered Species (CITES). The porbeagle (*Lamna nasus*), oceanic whitetip (*Carcharhinus longimanus*), scalloped hammerhead (*Sphyrna lewini*), smooth hammerhead (*Sphyrna zygaena*) and great hammerhead (*Sphyrna mokarran*) shark have also been proposed for listing in Appendix II of CITES due to the large volume and high value of their fins in the international trade market

It's curious to see genetic identification techniques criticized as, "relatively expensive and time consuming to employ". Such techniques (as we have seen) market themselves as quick, simple and efficient; yet Abercrombie et al. raise somewhat valid concern against their potential lack of accessibility, especially in the case of, "countries where resources available are limited." However, fin-based physical identification is by no means a replacement for genetic ID, as the latter's very existence emerged to address a classic (and, quite possibly, universal) problem: shark species are damn near impossible to identify with fins alone. Abercrombie et al. present their guide as preliminary step before genetic analysis, so that handlers could, "assess whether or not genetic testing is warranted (e.g., when fins from prohibited species are thought to be present)."

Carcharhinus leucas has been rarely described with uniquely specific physical details (we are by now all too familiar with the species' broad snout, teeth, fins, etc.). Yet, Abercrombie et al. provide something that we have never seen before: a precise description of texture. To both the first dorsal and pectoral fins, the authors attribute a rough, sandpaper-like texture composed of, "large, obvious denticles, some of which look like salt grains". This description is a bit like poetry, in the sense that it introduces sensation to our picture of *Carcharhinus leucas*. We understand the shark's shape, color, habits, cultural impact and elemental presence, but we have never before engaged with the species on tactile level. Salt and sandpaper are easy to visualize; can you imagine?

Place your right palm upon the Bull Shark's head. Trace your left index finger on the edge of the first dorsal fin. Look closely at the skin— gray, bronze, and shimmering in the sunlit ripples. Can you see the salt grains? Can you see the lazuli speckles? Can you feel the smoothness, the roughness—the shamanic dichotomy? Transform *understanding*.

2013

*Opportunistic Visitors: Long-Term Behavioural Response
of Bull Sharks to Food Provisioning in Fiji*
PLoS ONE, 8 (3): e58522
BRUNNSCHWEILER, J.M. & BARNETT, A.

Shark-based tourism that uses bait to reliably attract certain species to specific sites so that divers can view them is a growing industry globally, but remains a controversial issue. We evaluate multi-year (2004-2011) underwater visual (n=48 individuals) and acoustic tracking data (n=82 transmitters; array of up to 16 receivers) of bull sharks *Carcharhinus leucas* from a long-term shark feeding site at the Shark Reef Marine Reserve and reefs along the Beqa Channel on the southern coast of Viti Levu, Fiji. Individual *C. leucas* showed varying degrees of site fidelity. Determined from acoustic tagging, the majority of *C. leucas* had site fidelity indexes >0.5 for the marine reserve (including the feeding site) and neighbouring reefs. However, during the time of the day (09:00—12:00) when feeding takes place, sharks mainly had site fidelity indexes <0.5 for the feeding site, regardless of feeding or non-feeding days. Site fidelity indexes determined by direct diver observation of sharks at the feeding site were lower compared to such values determined by acoustic tagging. The overall pattern for *C. leucas* is that, if present in the area, they are attracted to the feeding site regardless of whether feeding or non-feeding days, but they remain for longer periods of time (consecutive hours) on feeding days. The overall diel patterns in movement are for *C. leucas* to use the area around the feeding site in the morning before spreading out over Shark Reef throughout the day and dispersing over the entire array at night.

It appears that Shark Reef Marine Reserve (SRMR)'s local feeding operations do not significantly disrupt the natural ecology of *Carcharhinus leucas*. While the 10-year old SCUBA ecotourism program has quite possibly attracted a greater number of Bull Sharks to the area over the course of existence, it has by no means altered the species' 'human-adverse' temperament; as demonstrated by Brunnschweiler & Barnett, "*C. leucas* avoid the area when humans are present, and hence food provisioning is essential to elicit human-oriented *C. leucas* behavior." On a larger scale, the SRMR's feeding operation, "does not appear to drive their long-term movements and the sharks are not strongly conditioned as otherwise they would be expected to be present at almost every feed."

Critics of such ecotourism programs worry about their potential to attract unnaturally high numbers of sharks to a specific area, and thereby increase their susceptibility to local fishing (or poaching, as in the case of SRMR). Champions of such programs counter with the fact that ecotourism improves conservation awareness, and furthermore financially secures the operation of protected refuges like Shark Reef Marine Reserve (which is an excellent and internationally renowned showcase of *Carcharhinus leucas* in its natural setting). It appears that this particular study supports the logic of the latter: as Bull Sharks do not seem to 'concentrate' outside of their natural tendencies, their risk of poaching is probably minimal—and certainly pales in comparison to the enormous benefits of the Shark Reef Marine Reserve's ecotourism-based protection.

Continued monitoring of the SRMR is warranted, however, as the program is very young, and the criticisms are mindful. Currently, *Carcharhinus leucas* leaves the SRMR area every year—perhaps retreating to a (as of yet unknown) pupping site. Any change in this or other natural behaviors (like diel movements) should command further inquiry.

2013

Use of Human-Altered Habitats by Bull Sharks in a
Florida Nursery Area
Marine and Coastal Fisheries: Dynamics, Management,
and Ecosystem Science, 5 (1): 28-38
CURTIS, T.H. & PARKYN, D.C. & BURGESS, G.H.

Bull Sharks *Carcharhinus leucas* in the Indian River Lagoon,
Florida, have been documented to frequently occur in human-altered
habitats, including dredged creeks and channels, boat marinas, and power
plant outfalls. The purpose of this study was to examine the short-term
movements of age-0 and juvenile Bull Sharks to quantify the extent to
which those movements occur in altered habitats. A total of 16 short-term
active acoustic tracks (2—26h) were carried out with 9 individuals, and a
10th individual was fitted with a long-term coded transmitter for passive
monitoring by fixed listening stations. Movement and activity space
statistics indicated high levels of area reuse over the span of tracking (hours
to days). All but one shark used altered habitat at some point during
tracking, such that 51% of all tracking positions occurred in some type of
altered habitat. Of the sharks that used altered habitat, the mean (± 1 SD)
percent of positions within altered habitat was 66 (± 40)%. Furthermore,
tracks for 3 individuals indicated selection for altered habitats. The single
passively monitored Bull Shark was detected in power plant outfalls almost
daily over a 5-month period, providing the first indication of longer-term
fidelity to thermal effluents. Use of one dredged creek was influenced by
local salinity, the tracked sharks dispersing from the altered habitat when
salinity declined.

300

Affinity to altered habitat is a serious problem. *Carcharhinus leucas* is a coastal species with a tendency towards site fidelity (especially when younger): in this sense, it is a species *without choice* in its surroundings—a species forced to endure the frontlines of coastal development and all of its extensive damage (including the perilous infusion of contaminants into the water). Indian River Lagoon is of particular concern as the entire system can be considered "altered". Local human activity has contributed to nutrient enrichment, reduced water quality, contamination, extirpation of the Smalltooth Sawfish (*Pristis pectinata*), and damage to seagrass, saltmarsh, and mangrove habitat.

In place of these natural systems are dredged creeks and channels (including the Intracoastal Waterway), marinas, boat ramps, causeways and power plant outfalls. The thermal activity of the latter can, "concentrate prey, especially during colder periods, when a variety of species use the effluents as thermal refugia." This phenomenon may be a chief driver of concentrated Bull Shark activity around power plants; according to Curtis et al, "If Bull Shark movements and habitat use reflect optimal foraging (energy maximization) strategies, then they will presumably select habitats with the highest prey densities." This logic well-explains increased Bull Shark activity around the Crane Creek boast basin, as the complex, "provides structure and refuge for prey species such as mullet *Mugil* spp., Hardhead Catfish *Ariopsis felis*, Gafftopsail Catfish *Bagre marinus*, and other fishes, confining them to an area of approximately 200 x 200 m."

Exposure to altered habitat (and pursuit of 'altered prey') has resulted in the bioaccumulation of mercury and polychlorinated biphenyls within the Bull Sharks of the Indian River Lagoon. Tissues from these sharks exceed FDA-safe mercury levels and contain, "among the highest contamination levels of any Florida marine species tested."

2013

Comparative Feeding Ecology of Bull Sharks
(Carcharhinus leucas) in the Coastal Waters of the SW
Indian Ocean Inferred from Stable Isotope Analysis
PLoS ONE, 8 (10): e78229
DALY, R. & FRONEMAN, P.W. & SMALE, M.J.

As apex predators, sharks play an important role shaping their respective marine communities through predation and associated risk effects. Understanding the predatory dynamics of sharks within communities is, therefore, necessary to establish effective ecologically based conservation strategies. We employed non-lethal sampling methods to investigate the feeding ecology of bull sharks (*Carcharhinus leucas*) using stable isotope analysis within a subtropical marine community in the southwest Indian Ocean. The main objectives of this study were to investigate and compare the predatory role that sub-adult and adult bull sharks play within a top predatory teleost fish community. Bull sharks had significantly broader niche widths compared to top predatory teleost assemblages with a wide and relatively enriched range of $\delta^{13}C$ values relative to the local marine community. This suggests that bull sharks forage from a more diverse range of $\delta^{13}C$ sources over a wider geographical range than the predatory teleost community. Adult bull sharks appeared to exhibit a shift towards consistently higher trophic level prey from an expanded foraging range compared to sub-adults, possibly due to increased mobility linked with size. Although predatory teleost fish are also capable of substantial migrations, bull sharks may have the ability to exploit a more diverse range of habitats and appeared to prey on a wider diversity of larger prey.

Large sharks like *Carcharhinus leucas* possess a unique ecological power that most predatory teleosts cannot replicate. This is the power of trophic unification—the ability to connect spatially separated communities by means of common energy transfer (upwards to the foraging apex predator). One adult Bull Shark can forage within many disparate systems thanks to its enhanced size, mobility and appetite (juvenile Bull Sharks— and many predatory teleosts—are found to be lacking in all three attributes). It is beyond the margins of juvenile and teleost capability that the adult *Carcharhinus leucas* thrives; by exploiting such a massive generalist niche across disparate systems, a full-grown Bull Shark exerts a powerful trophic influence and, "ensures energy transfer that link ecological processes which may maintain the functionality of these systems."

On the east coast of southern Africa, adult Bull Sharks seamlessly transition between coastal and pelagic food webs, and consume fish species from four trophic groups (as identified by Daly et al.). Group 3 (coastal consumers) was found to constitute the largest proportion of potential Bull Shark diet, and is composed of coastal and benthic mesopredators such as the Humpback Snapper (*Lutjanus gibbus*), Slender Baardman (*Umbrina robinsoni*) and Natal Stumpnose (*Rhabdosargus sarba*). Group 2 (top coastal predators) and Group 1 (top pelagic predators) become increasingly important to the Bull Shark as it grows; increased mobility and predatory ability enables adult *Carcharhinus leucas* to access this piscivorous conglomerate of more specialized hunters such as the Dorado (*Coryphaena hippurus*), Sailfish (*Istiophorus platpterus*), and King Mackerel (*Scomberomorus commerson*). Group 4 is composed of two elasmobranch species: *Carcharhinus limbtus* (Blacktip Shark) and *Rhizoprionodon acutus* (Milk Shark). Both species are accessible to fully realized Bull Sharks, and are likewise generalist predators.

2013

Multiscale Analysis of Factors that Affect the Distribution of Sharks throughout the Northern Gulf of Mexico
Fishery Bulletin, 111 (4): 370-380
DRYMON, J.M. & CARASSOU, L. & POWERS, S.P. &
GRACE, M. & DINDO, J. & DZWONKOWSKI, B.

Identification of the spatial scale at which marine communities are organized is critical to proper management, yet this is particularly difficult to determine for highly migratory species like sharks. We used shark catch data collected during 2006−09 from fishery-independent bottom-longline surveys, as well as biotic and abiotic explanatory data to identify the factors that affect the distribution of coastal sharks at 2 spatial scales in the northern Gulf of Mexico. Centered principal component analyses (PCAs) were used to visualize the patterns that characterize shark distributions at small (Alabama and Mississippi coast) and large (northern Gulf of Mexico) spatial scales. Environmental data on temperature, salinity, dissolved oxygen (DO), depth, fish and crustacean biomass, and chlorophyll-a (chl-a) concentration were analyzed with normed PCAs at both spatial scales. The relationships between values of shark catch per unit effort (CPUE) and environmental factors were then analyzed at each scale with co-inertia analysis (COIA). Results from COIA indicated that the degree of agreement between the structure of the environmental and shark data sets was relatively higher at the small spatial scale than at the large one. CPUE of Blacktip Shark (*Carcharhinus limbatus*) was related positively with crustacean biomass at both spatial scales. Similarly, CPUE of Atlantic Sharpnose Shark (*Rhizoprionodon terraenovae*) was related positively with chl-a concentration and negatively with DO at both spatial scales.

304

While *Carcharhinus leucas* was formally incorporated into Drymon et al.'s compelling analysis, its numbers from both the large-scale and small-scale datasets were (unfortunately) too low to yield any significant patterns within the PCA and COIA. Forty Bull Sharks (ranging in size from 73.0—155.5 cm FL) were captured within the small-scale survey (pertaining to the Alabama and Mississippi coasts); mean CPUE was 0.11 (±0.02) sharks per 100 hooks per hour. Only twenty-one Bull Sharks (ranging in size from 131.4—176.0 cm FL) were captured within the large-scale survey (pertaining to the northern Gulf of Mexico); mean CPUE was 0.06 (±0.01) sharks per 100 hooks per hour.

It's a shame to lack a species-specific understanding of *Carcharhinus leucas* in accordance with Drymon et al.'s analyses, as other species demonstrated fascinating distributional preferences. Blacktip Sharks (*Carcharhinus limbatus*) strongly corresponded with crustacean biomass at both the large and small survey scales. This correlation may not be prey-related (as the species is chiefly piscivorous), but its explanation is cryptic. The distribution of Atlantic Sharpnose Sharks (*Rhizoprionodon terraenovae*) corresponded strongly with chl-*a* concentration. This finding may relate to the species' trophic plasticity, as the sharks may be selecting for areas of general productivity (as indicated by chl-*a*) rather than for high concentrations of specific prey items. Both the Atlantic Sharpnose Shark and the Spinner Shark (*Carcharhinus brevipinna*) demonstrated a negative distributional correlation with dissolved oxygen; Drymon et al. specifically relate this observation to Heithaus et al.'s 2009 findings on juvenile *Carcharhinus leucas*, stating that, "although dissolved oxygen is not as widely considered as temperature or salinity, it may play an important role as an environmental influence that affects the distribution of top predators in coastal environments."

2013

Sharks, Batoids, and Chimaeras of the North Atlantic
**FAO Species Catalogue for Fishery Purposes, 7. Rome,
FAO: 523 pp.**
EBERT, D.A. & STEHMANN, M.

This volume is a comprehensive, fully illustrated Catalogue of the
Sharks, Batoid Fishes, and Chimaeras of the North Atlantic, encompassing
FAO Fishing Areas 21 and 27. The present volume includes 11 orders, 32
families, 66 genera, and 148 species of cartilaginous fishes occurring in the
North Atlantic. The Catalogue includes a section on standard measurements
for a shark, batoid, and chimaera, with associated terms. It provides
accounts for all orders, families, and genera and all keys to taxa are fully
illustrated. Information under each species account includes: valid modern
names and original citation of the species; synonyms; the English, French,
and Spanish FAO names for the species; a lateral view for sharks and
chimaeras, dorsal and often also ventral view for batoids, and often other
useful illustrations; field marks; diagnostic features; distribution, including
a GIS map; habitat; biology; size; interest to fisheries and human impact;
local names when available; a remarks sections; and literature. The volume
is fully indexed and also includes sections on terminology and
measurements including an extensive glossary, a list of species by FAO
Statistical Areas, and a dedicated bibliography.

This document was prepared under the general supervision of the
FishFinder Programme of the Marine Resurces Service, Fisheries
Resources and Environment Division, Fisheries Department, Food and
Agriculture Organization of the United Nations (FAO), Rome, Italy.

This FAO publication is brimming with new information on *Carcharhinus leucas*. The previous estimated maximum size of 350 cm is shattered by a new (and unexpected) theoretical: a whopping 400 cm maximum (based on unconfirmed reports). Size at maturity is estimated to be 157 to 226 cm for males and 180 to 230 cm for females, while size at birth for ranges from 56 to 81 cm. Growth rates vary with age (and are as follows):

Age Bracket (in years)	Growth Rate
0—5	15—20 cm / yr.
6—10	10 cm / yr.
11—16	5—7 cm / yr.
Maturity (M: 14 to 15) (F: 18)	4—5 cm / yr.

Maximum age for *Carcharhinus leucas* is alleged to be, "about 21 to 32 years for males and 24 to 29 years for females, but with an estimated longevity of about 50 years for some populations." In captivity (specifically, public aquaria), Bull Sharks have been successfully maintained for over 20 years.

Bull Sharks of the Western Atlantic mate off the coast of Florida and within the Gulf of Mexico throughout June and July. Courtship is aggressive, and fighting scars can be found on both sexes (though females are often more scarred than males). After a gestation period of 10 to 12 months, pups are born within estuaries and river mouths around late spring and summer. Ideal estuarine and riverine habitat would consist of high freshwater inflow. Though Bull Sharks exhibit high site fidelity, they are migratory, and typically, "travel an average of 5 to 6 km per day and spend most of their time in water less than 20 m deep where the water temperature is on average 26°C or more."

2013

An Annotated Checklist of the Chondrichthyans of Taiwan
Zootaxa, 3752: 279-386
EBERT, D.A. & WHITE, W.T. & HO, H.-C. & LAST, P.R. &
NAKAYA, K. & SÉRET, B. & STRAUBE, N. & NAYLOR,
G.J.P. & DE CARVALHO, M.R.

An annotated checklist of chondrichthyan fishes (sharks, batoids, and chimaeras) occurring in Taiwanese waters is presented. The checklist is the result of a biodiversity workshop held in March 2012 as well as on-going systematic revisions by the authors. The chondrichthyan fauna of Taiwan is one of the richest in the world with the number of species totaling 181, comprising 52 families and 98 genera. It includes 31 families, 64 genera, and 119 species of sharks, 19 families, 31 genera, and 58 species of batoids, and 2 families, 3 genera, and 4 species of chimaeras. The most species-rich families are the Carcharhinidae with 22 species followed by the Scyliorhinidae with 17. The most species-rich batoid families are the Dasyatidae with 11 species and and the Rajidae with 10. Verified voucher material is provided for each species where available and potential taxonomic issues are high-lighted when applicable. This represents the first detailed, evidence-based checklist of chondrichhtyans from Taiwanese waters in over 40 years.

Carcharhinus leucas (Müller & Henle, 1839)

Bull Shark /低鰭真鯊

Carcharias (Prionodon) leucas Müller & Henle, 1839; 42. Syntypes (4): only 2 stuffed syntypes still in existence; MNHN A-9650, adult male 1600 mm, MNHM A-9652, female 1869 mm, Antilles, Western Atlantic.

C. leucas is rare in Taiwanese waters, and no local specimens were examined as voucher material for this checklist. Interestingly, Ebert et al. make specific note of Naylor et al.'s 2012 molecular study, stating that, "this species may represent a species complex, which is in need of urgent taxonomic research." Only IUCN Conservation status (NT) and local synonymy (*Carcharhinus leucas*: Chen & Yu, 1986; Chen & Joung, 1993; Shen & Wu, 2011) are mentioned in addition to these remarks.

Taiwanese Bull Sharks may be part of the Bornean subset of the species complex, *Carcharhinus* cf. *leucas* 1. If this is a truly separate species, then all previously covered material on Bull Sharks adjacent to the Bornean system (such as The Legend of Surabaya and the ecology of Mekong Bulls) is misapplied to the true *C. leucas*, and must instead be reallocated to an entirely new identity—an identity beyond our reach.

Across time and culture, we have aspired to *understand* the Bull Shark as it *is*. The true identity of this species is present and unchanging: it exists! At this very moment, crystalline to the cosmos, the Bull Shark *is*— an entity that does not bother with human calculation and convention: independent of the human eye (which is imperfect in its discernment), the true species that we call *Carcharhinus leucas* exists in perfect identity. We are naturally alien to that truth, and yet, we adhere to a magnificent aspiration: to *understand* that perfect identity beyond our innateness. This quest for *understanding* has been riddled with human error, and thus, the true form of *Carcharhinus leucas* dances beyond our grasp—leaping from possibility to possibility with enticing alacrity.

Such possibilities tease us now. Is the true *Carcharhinus leucas* the same that crowns Surabaya? Does the true *Carcharhinus leucas* lurk in the Gold Coast canals? Or is the true form a complete breach of cosmopolitism—an authentically Atlantic species?

2013

Trace and Minor Element Chemistry of Modern Shark Teeth and Implications for Shark Tooth Geochronometry
Abstract

Geological Society of America Abstracts with Programs
45 (2): 72

JOHN, J. & CHAMBERLAIN, J.A. & TAPPERO, R.

The hard parts of marine organisms reflect the elemental and isotopic composition of the seawater in which they form. We have determined the average concentrations of the following trace and minor elements in teeth from a variety of coastal and pelagic shark species: Fe, Cr, Co, Mn, Ba, Cu, Al, U, Pb, Ni, V, Zn, Mg, Ce, and Sr. The teeth analyzed were collected from the jaws of recently deceased individuals of the following species: *Galeocerdo cuvier* (tiger shark), *Carcharhinus limbatus* (black tip), *Carcharias taurus* (sand tiger), *Carcharhinus leucas* (bull shark), *Prionace glauca* (blue shark), *Isurus oxyrinchus* (mako shark), *Carcharhinus brevipinna* (spinner shark), *Carcharhinus obscurus* (dusky shark), and *Hexanchus griseus* (bluntnose six gill shark). Trace and minor element concentrations were measured using an ICP-MS; tooth enameloid ranged from <1 ppm for U, to 1000s of ppm for Sr. For individual sharks elemental concentrations in enameloid did not vary greatly from tooth to tooth in the functional tooth row, nor was there wide variation in concentration between teeth in functional and adjacent pre-functional tooth rows. The concentrations for each particular element in enameloid were generally similar from species to species, with some notable exceptions (e.g., relatively low Sr in *Carcharhinus brevipinna*).

As this is only an abstract to a live presentation, we lack specific details regarding the concentrations of elements comprising *Carcharhinus leucas* teeth; however, the enameloid of all nine shark species examined by the authors is generally uniform in chemical structure. John et al. specifically interpret this observation to mean that, "the life habits of the animals we tested and their food sources were sufficiently broad to have exposed our sharks to average conditions of oceanic chemistry." In understanding the elemental concentrations of modern shark teeth, John et al. provide, "base level values that allow for the detection of diagenetic changes in fossil shark tooth chemistry."

Each of the chemical elements that compose the cutting crowns of *Carcharhinus leucas* is metallic: three alkaline earth metals (Mg, Sr, and Ba), one lanthanoid (Ce), one actinoid (U), eight transition metals (V, Cr, Mn, Fe, Co, Ni, Cu, and Zn) and two post-transition metals (Al and Pb) contribute to the strength, sharpness and sheen of the Bull Shark's teeth. The proper names and atomic numbers (Z) of these metals are as follows:

Symbol	Z	Element	Symbol	Z	Element
Mg	12	Magnesium	Cu	29	Copper
Al	13	Aluminium	Zn	30	Zinc
V	23	Vanadium	Sr	38	Strontium
Cr	24	Chromium	Ba	56	Barium
Mn	25	Manganese	Ce	58	Cerium
Fe	26	Iron	Pb	82	Lead
Co	27	Cobalt	U	92	Uranium
Ni	28	Nickel			

2013

The Fishes of the Inland Waters of Southeast Asia: A Catalogue and Core Bibliography of the Fishes Known to Occur in Freshwaters, Mangroves and Estuaries
The Raffles Bulletin of Zoology, Supplement No. 27: 1-663
KOTTELAT, M.

There are 3108 valid and named native fish species in the inland waters of Southeast Asia between the Irrawaddy and Red River drainages, the small coastal drainages between the Red River and Hainan, the whole Indochinese Peninsula, Andaman and Nicobar Islands, Indonesia (excluding Papua Province, Waigeo, Aru [but Kai is included]), and the Philippines. They belong to 137 families. Their taxonomy and nomenclature are reviewed. The original descriptions of all 7047 recorded species-group names and 1980 genus-group names have been checked in the original works for correct spelling, types, type locality and bibliographic references. The bibliography includes about 4700 titles. Synonymies are given, based on published information as well as unpublished observations.

The names of 49 introduced species and 347 extralimital taxa cited in the discussions have also been checked. The original descriptions of all species not present in the covered area but cited as type species of genera have been checked for availability, authorship, date and correct spelling. The availability of some family-group names has been checked when there was suspicion of possible nomenclatural problems.

Bibliographic notes include new informations on the dates of publication of works by, among others, Bleeker, Bloch, Heckel and Steindachner and discussion of authorship of names in various works. The main nomenclatural acts are listed below:

Kottelat has some very definite opinions when it comes to the present state (and culture) of taxonomic nomenclature. Within his impressive volume, the author deliberately includes an introduction to the basic principles and terminology defining nomenclature, citing as his motivation a 'sad reality' that, "a majority of the users of scientific names, especially those in the geographic area covered by this list, have not had the opportunity to study the rules." Kottelat additionally claims that this introduction is likewise necessary for, "most researchers completing their studies in western countries."

Many errors (or erroneous approaches) have unfortunately manifested within this field. Common among them is the confusion between nomenclature and taxonomy: nomenclature deals only with the scientific name (binomial) itself, while taxonomy deals with the biological relationships inferred to exist between taxa. A similar confusion exists in regards to the role of type specimens (types) in nomenclature: as succinctly defined by Kottelat, the type specimen is, "the type of a name, not of a species." The four syntypes that define *Carcharhinus leucas* (Müller & Henle, 1839) are thus not 'models' of the species itself, but merely bearers of the species' legal name. Categories of types include:

Holotype	Sole specimen upon which the binomial (nominal species) is based.
Syntype	IF NO *holotype*: One of multiple specimens establishing the binomial.
Paratype	IF *holotype*: Additional specimen(s) that establish(es) the binomial.
Lectotype	IF NO *holotype*: a *syntype* (from a series) valued as a *holotype*.
Paralectotype	IF NO *holotype*: a *syntype* (from a series) devalued by the *lectotype*.
Neotype	IF *holotype*, *lectotype*, or type series LOST: then valued as a *holotype*.
Allotype	A *paratype* of a sex that differs from that of the *holotype*.
Topotype	A specimen collected from the locality of the primary type specimen.

2013

DNA Barcoding of Shark Meats Identify Species Composition and CITES-Listed Species from the Markets in Taiwan
PLoS ONE, 8 (11): e79373
LIU, S.-Y.V. & CHAN, C.-L.C. & LIN, O. & HU, C.-S. &
CHEN, C.A.

An increasing awareness of the vulnerability of sharks to exploitation by shark finning has contributed to a growing concern about an unsustainable shark fishery. Taiwan's fleet has the 4th largest shark catch in the world, accounting for almost 6% of the global figures. Revealing the diversity of sharks consumed by Taiwanese is important in designing conservation plans. However, fins make up les than 5% of the total body weight of a shark, and their bodies are sold as filets in the market, making it difficult or impossible to identify species using morphological traits.

In the present study, we adopted a DNA barcoding technique using a 391-bp fragment of the mitochondrial cytochrome oxidase I (COI) gene to examine the diversity of shark filets and fins collected from markets and resturants island-wide in Taiwan.

Amongst the 548 tissue samples collected and sequenced, 20 major clusters were apparent by phylogenetic analyses, each of them containing individuals belonging to the same species (most with more than 95% bootstrap values), corresponding to 20 species of sharks. Additionally, *A. pelagicus*, *C. falciformis*, *I. oxyrinchus*, and *P. glauca* consisted of 80% of the samples we collected, indicating that these species might be heavily consumed in Taiwan. Approximately 5% of the tissues samples used in this study were identified as species listed in CITES Appendix II, including two species of *Sphyrna*, *C. longimanus* and *C. carcharias*.

Only two samples of *Carcharhinus leucas* were identified in this study. From the entire collection 548 COI gene sequences (corresponding to 20 different shark species), Liu et al. generated two phylogenetic models: a neighbor-joining tree and a maximum-likelihood tree (with *Rhinoptera steindacheri* as the outgroup). Both trees placed *Carcharhinus leucas* most closely with *Carcharhinus limbatus* (in agreement with most previous studies). However, each tree simultaneously generated a very different arrangement for the Carcharhiniformes as a whole. In the neighbor-joining tree, *Prionace glauca* was placed within the genus *Carcharhinus*, while the genus *Sphyrna* and species *Scoloidon laticaudus* and *Galeocerdo cuvier* were posited as, "some of the first lineages to branch off at the base of the Carcharhiniform clade." In the maximum-likelihood tree, *Sphyrna* and *Galeocerdo cuvier* were placed within the genus *Carcharhinus* instead, and *Prionace glacua* was alternatively posited as, "sister to the *Carcharhinus* clade".

Carcharhinus leucas may be rare in Taiwan, but it is part of a superlative assemblage, as Taiwan has, "the highest species diversity of sharks in the world." While Taiwan is the first Asian nation to ban shark finning (since 2012), its fleet has, "the 4th largest shark catch in the world, with a declared 6 million sharks caught annually, accounting for almost 6% of the global figures." According to Liu et al., these numbers could be, "greatly underestimated". Capture and trade of CITES Appendix II species like *Carcharhinus longimanus*, *Carcharodon carcharias*, *Sphyrna lewini* and *Sphyrna zygaena* is of grave concern, especially in consideration of the fact that all but the White Shark *(Carcharodon carcharias)* constitute a significant portion of Taiwanese landings. As DNA barcoding (using COI) effectively identifies processed sharks to the species level, it could be a powerful management tool for the tracking of at-risk species trade.

2013

Cestodes of the Bull Shark Carcharhinus leucas in Chachalacas Beach, Veracruz, Mexico

Neotropical Helminthology, 7 (1): 167 - 171

MÉNDEZ, O. & GONZÁLEZ, M.A.D.

The spiral valve (intestine) of a bull shark *Carcharhinus leucas* (Valenciennes, 1841) caught by artisanal fisheries at Chachalacas Beach, Veracruz Mexico was examined for helminth parasites. A total of 303 collected cestodes belonging to eight species: *Callitetrarhynchus* cf. *gracilis* (Rudolphi, 1819); *Cathetocephalus* cf. *thatcheri* Dailey & Overstreet, 1973; Eutetrarhynchidae sp.; *Otobothrium* sp. 1; *Otobothrium* sp. 2; *Paraorygmatobothrium* sp. 1; *Paraorygmatobothrium* sp. 2 and *Phoreiobothrium* sp. The bull shark feeds on a wide range of prey mainly bony fish. This is the first record of cestodes in the bull shark off the coast of Veracruz Mexico.

La válvula espiral (intestine) de un tiburón toro *Carcharhinus leucas* (Valenciennes, 1841) capturado en la pesca artisanal de playa Chachalacas, Veracruz México fue examinado en busca de helmintos parásitos. Se colectaron 303 céstodos pertenecientes a ocho especies: *Callitetrarhynchus* cf. *gracilis*; *Cathetocephalus* cf. *thatcheri*; Eutetrarhynchidae sp.; *Otobothrium* sp. 1; *Otobothrium* sp. 2; *Paraorygmatobothrium* sp. 1; *Paraorygmatobothrium* sp. 2 y *Phoreiobothrium* sp. El tiburón toro se alimenta de una amplia gama de presas principalmente peces óseos, llegando a albergar un gran número de especies de cestodos. Este es el primer registro de céstodos en el tiburón toro par alas costas de Veracruz México.

The spiral valve is one of the most magnificent adaptations to evolve within the Selachii and few other fish (such as the rays, sturgeon, paddlefish, bichirs and lungfish). In essence, the spiral valve is a twisted, coiled infold of the shark's intestine: this design drastically increases intestinal surface area, and allows for the maximum absorption of nutrients passing through the shark's digestive system. It is through this miraculous anatomy that the Bull Shark can endure long periods without food. In a Romantic sense, the spiral valve is the epitome of frugality: no food goes to waste. The prudent Bull—a shark that will not allow such gifts to be squandered—savors each piece of prey.

Tapeworms are the antithesis of the spiral valve and its entire purpose. Buried within the intestine's many folds, these thieving parasites destroy the valve's efficiency and absorb what nutrients they can for themselves. About thirty species are known to plague *Carcharhinus leucas* with their incessant hunger. Of the eight species recorded by Méndez & González, six can be found in *Carcharhinus limbatus*: the Blacktip Shark. This finding may suggest that the Mexican Bull and Blacktip sharks occupy similar trophic niches and prey upon similar same-vector teleosts in the Gulf. Mississippian drums are briefly mentioned as a group of particular interest (as they are preyed upon by both shark species and serve as an intermediate host for the larval tapeworms).

The taxonomy of local tapeworms is poorly understood, and this record (while novel) applies to only one individual *Carcharhinus leucas*: what would an analysis of other Bull Sharks from this region yield? Méndez & González demonstrate an excellent potentiality with tapeworm parasites in that they could elucidate local trophic dynamics: would further analysis link the dietary pattern of *Carcharhinus leucas* more closely to that of *C. limbatus*, or would new worms simply reveal twists in the niche?

2013

Variable Delta N-15 Diet-Tissue Discrimination Factors among Sharks: Implications for Trophic Position, Diet and Food Web Models

PLoS ONE, 8 (10): e77567

OLIN, J.A. & HUSSEY, N.E. & GRGICAK-MANNION, A. & FRITTS, M.W. & WINTER, S.P. & FISK, A.T.

The application of stable isotopes to characterize the complexities of a species foraging behavior and trophic relationships is dependent on assumptions of $\delta^{15}N$ diet-tissue discrimination factors ($\Delta^{15}N$). As $\Delta^{15}N$ values have been experimentally shown to vary amongst consumers, tissues and diet composition, resolving appropriate species-specific $\Delta^{15}N$ values can be complex. Given the logistical and ethical challenges of controlled feeding experiments for determining $\Delta^{15}N$ values for large and/or endangered species, our objective was to conduct an assessment of a range of reported $\Delta^{15}N$ values that can hypothetically serve as surrogates for describing the predator-prey relationships of four shark species that feed on prey from different trophic levels (i.e., different mean $\delta^{15}N$ dietary values). Overall, the most suitable species-specific $\Delta^{15}N$ values decreased with increasing dietary- $\delta^{15}N$ values based on stable isotope Bayesian ellipse overlap estimates of shark and the principal prey functional groups contributing to the diet determined from stomach content analyses. Thus, a single $\Delta^{15}N$ value was not supported for this speciose group of marine predatory fishes.

Within their complex dietary analysis, Olin et al. generate a fascinating statistical summary of the functional prey groups (and principal prey) of *Carcharhinus leucas*. This information is derived from a combination of direct sampling (of the stomach contents from the South African Bull Sharks of KwaZulu-Natal) and consultation with published literature on the subject by Cliff & Dudley (1991) [9]. Four commonly applied dietary indices are utilized within this assessment: %F, the frequency of occurrence of each prey item; %N, the numbers of each prey item; %W, the weight contribution of a prey item, and; %IRI, the relative importance of the prey item (wherein IRI = %F[%N + %W]).

	%F	%N	%W	%IRI
MAMMAL	**7.7**	**2.6**	**7.5**	**0.6**
Cetacea	4.3	1.4	5.1	3.3[d]
ELASMOBRANCH	**57.5**	**20.7**	**65.5**	**40.6**
Guitarfish (Rhinobatidae)	9.5	3.2	17.8	11.4[c]
Dusky/Sharpnose shark (Carcharhinidae)	6.2	2.3	8	3.3[a]
Bull/Blue/Honeycomb stingray (Dasyatidae)	1.5	0.6	2.2	0.2[b]
TELEOST	**71.1**	**74.5**	**25.4**	**58.5**
Spotted grunter (*Pomadasys commersonnii*)	4.8	1.7	.6	1.9[e]
CRUSTACEAN	**5.3**	**1.8**	**0.1**	**0.1**
MOLLUSK	**5.8**	**2.0**	**3.8**	**0.1**

According to the authors, a $\Delta^{15}N$ of 2.3ppt provided the highest measure of overlap between predator and functional prey group ellipses for the Bull Shark (ellipses were generated from a Bayesian approach following a 2011 study by Jackson et al.[11])

2013

Gulf-Wide Decreases in the Size of Large Coastal Sharks
Documented by Generations of Fishermen
Marine and Coastal Fisheries: Dynamics, Management,
and Ecosystem Science, 5 (1): 93-102
POWERS, S.P. & FODRIE, F.J. & SCYPHERS, S.B. &
DRYMON, J.M. & SHIPP, R.L. & STUNZ, G.W.

Large sharks are top predators in most coastal and marine ecosystems throughout the world, and evidence of their reduced prominence in marine ecosystems has been a serious concern for fisheries and ecosystem management. Unfortunately, quantitative data to document the extent, timing, and consequences of changes in shark populations are scarce, thwarting examination of long-term (decadal, century) trends, and reconstructions based on incomplete data sets have been the subject of debate. Absence of quantitative descriptors of past ecological conditions is a generic problem facing many fields of science but is particularly troublesome for fisheries scientists who must develop specific targets for restoration. We were able to use quantitative measurements of shark sizes collected annually and independently of any scientific survey by thousands of recreational fishermen over the last century to document decreases in the size of large sharks from the northern Gulf of Mexico. Based on records from fishing rodeos in three U.S. coastal states, the size (weight or length) of large sharks captured by fishermen decreased by 50-70% during the 20 years after the 1980s. The pattern is largely driven by reductions in the occurrence and sizes of Tiger Sharks *Galeocerdo cuvier* and Bull Sharks *Carcharhinus leucas* and to a lesser extent Hammerheads *Sphyrna* spp.

Powers et al. brilliantly incorporate the knowledge of local fishermen from three tournaments (the Alabama Deep-Sea Fishing Rodeo, Mississippi Deep-Sea Fishing Rodeo, and Texas Deep-Sea Roundup) to assist in the elucidation of shark populational trends in the northern Gulf of Mexico. Though these extensive records span a total of 80 years, they do not meet most empirical standards; however, they are nevertheless essential glimpses of a disturbing pattern: a post-1980's Gulf-wide disappearance of prize-winning Bull, Tiger and other large predatory sharks.

All three tournaments reported the same trend: as put by the authors, "the average size of the three largest sharks captured in all three rodeos increased from the rodeos' inceptions until the early 1980s, but decreased by >60% in the late 1980s and remained low through 2009, the last year included in the data set". The initial increases in reported size can be most likely explained by increases in fishing technology and communication (which themselves lead to a general improvement in fishing efficiency). In tandem with this recreational improvement was the commercial improvement of the shark longlines (which peaked in the Gulf by the late 1980s, but thereafter experienced sharp declines). The combination of recreational and commercial declines following the 1980s — even in the face of improved fishing efficiency —strongly suggests that large sharks may have *themselves* become less abundant in the Gulf of Mexico as a result of overexploitation: the prize-winning Bull Sharks of old may have vanished from the Gulf coast entirely.

One of the most disturbing aspects of this report is its reflection of the "shifting baseline" theory: most younger fishermen (lacking the experience of older generations) did not report a decline in shark sizes, whereas most older fishermen (over 60) generally affirmed the decline, and therefore, "viewed changes in size of sharks more accurately".

2013

The Role of Non-Resident Sharks in Shaping
Coral Reef Communities Abstract
9th Indo-Pacific Fish Conference (IPFC), Abstracts: 148
SIMPFENDORFER, C.A. & ESPINOZA, M. & HEUPEL, M.R.
& TOBIN, A.J.

The importance of sharks in shaping coral reef communities, and the effects of declines in reef shark numbers, has received increased attention in recent times. Relevant research has focused on reef-resident species of shark, most notably grey reef (*Carcharhinus amblyrhynchos*), whitetip reef (*Triaenodon obesus*) and blacktip reef (*Carcharhinus melanopterus*) sharks. However, a range of other shark species occur around and within coral reefs, and may play an important role as predators of resident reef invertebrates, fish and sharks. This predation may have both direct and indirect effects on the functioning of coral reef communities. Catch data from reefs in the southern Great Barrier Reef indicates that species such as tiger (*Galeocerdo cuvier*), sandbar (*Carcharhinus plumbeus*), blacktip (*Carcharhinus limbatus*), bull (*Carcharhinus leucas*) and weasel (*Hemigaleus australiensis*) sharks regularly occupy coral reef ecosystems in which they may play important roles. Acoustic monitoring data that show the space use of non-resident species will be presented. The results of this preliminary work indicates that the role of non-resident shark species is far greater than previously anticipated, and may be an important consideration in the management of coral reef ecosystems.

Simpfendorfer et al. identify four subcategories for the 'non-resident' group of sharks that visit the Great Barrier Reef: individual specialists (species that exhibit high site fidelity to the reef but additionally utilize a wide range of habitats outside of the reef area); nocturnal invaders (species that only enter the reef at night); seasonal invaders (species that seasonally migrate over long distances into the reef), and; occasional invaders (species that interact with the reef community only sparingly). *Carcharhinus leucas* is a seasonal invader; at one point, Simpendorfer et al. tracked eight individuals from New South Wales into the Great Barrier Reef. It is speculated that Bull Sharks form an important part of the long-distance energy flow between these two communities.

Bull Sharks have a greater impact on the reef than occasional invaders, but their inconsistent presence renders them less influential than nocturnal invaders or individual specialists. Exemplary shark species of each 'non-resident' category are provided below:

Category	Common Name	Scientific Name
Seasonal Invader	Bull Shark	*Carcharhinus leucas*
Seasonal Invader	Sandbar Shark	*Carcharhinus plumbeus*
Individual Specialist	Tiger Shark	*Galeocerdo cuvier*
Individual Specialist	Lemon Shark	*Negaprion brevirostris*
Individual Specialist	Australian Weasel Shark	*Hemigaleus australiensis*
Occasional Invader	Scalloped Hammerhead	*Sphyrna lewini*
Occasional Invader	Great Hammerhead	*Sphyrna mokarran*
Occasional Invader	Common Blacktip Shark	*Carcharhinus limbatus*
Nocturnal Invader	Sliteye Shark	*Loxodon macrorhinus*

2013

The Coastal Fishes of the Cape Verde Islands — New Records and an Annotated Check-List (Pisces)

Spixiana, 36 (1): 113-142

WIRTZ, P. & BRITO, A. & FALCÓN, J.M. & FREITAS, R. & FRICKE, R. & MONTEIRO, V. & REINER, F. & TARICHE, O.

A check-list of the coastal fishes of the Cape Verde Islands is presented. The species *Acantholabrus palloni*, *Canthigaster supramacula*, *Carcharhinus leucas*, *Chaetodipterus lippie*, *Corniger spinosus*, *Dasyatis centroura*, *Didogobius* n sp., *Epigonus constanciae*, *Halobatrachus didactylus*, *Hemiramphus balao*, *Leptocharias smithii*, *Lobotes surinamensis*, *Malacoctenus* n. sp., *Megalops atlanticus*, *Mugil bananensis*, *Mugil capurrii*, *Negaprion brevirostris*, *Rhinecanthus aculeatus*, *Sardinella aurita*, *Sciaena umbra*, *Serranus heterurus*, *Sphyraena barracuda*, *Uranoscopus cadenati*, and *Zu cristatus* are recorded for the first time from the Cape Verde Islands. We have recognized 77 previous records as identification errors or registration errors and indicate 35 other records as doubtful. Including the 24 new records we now list 315 fish species from the coastal waters of the Cape Verde Islands. Twenty of them (6.3%) appear to be endemic to the archipelago.

Carcharhinus leucas (Müller & Henle [ex Valenciennes] 1839)
References: First record.
Remarks: Recorded as moderately common in an unpublished report of the INDP by Vera Gominho; this appears to be the first real evidence for the Cape Verde Islands.

324

Two major currents dictate the character of the Cape Verde Archipelago and its marine life: the North Equatorial Counter-Current (NECC, which flows eastward from French Guiana in South America to Sierra Leone in Africa) and the Canary Current (CC, which flows southwestward from Morocco to Senegal, and eventually into the NECC offshore). The NECC covers the entire Cape Verde Archipelago in July, but gradually shifts southeastward as the season progresses (until eventually reaching its lowest point southeast of the Cape Verde Islands from October to March). As the NECC expands northwestward into the Archipelago from April to June, the CC simultaneously flows into Cape Verde from the northeast (at its widest span and greatest intensity). After July, the CC weakens, narrows, and adheres much more closely to the African mainland.

This complex exchange of currents infuses Cape Verde with a distinctly tropical character (much in juxtaposition to Senegal's comparatively temperate waters at the same latitude). Cold upwelling along the African coast contribute to the mainland's cooler winter surface water temperatures (reaching as low as 14°C). Cape Verde's winter water temperatures rarely drop below 20°C (most likely as a result of not having this upwelling). These stark abiotic differences lend to Cape Verde's being more similar in faunal composition to the tropical Gulf of Guinea than the northwestern coast of Africa.

Twenty species of coastal fish (defined as species found within the 60m depth contour) are thought to be endemic to Cape Verde. While Cape Verde's isolation from the African mainland greatly contributes to this high degree of endemism, larger fish species (like *Carcharhinus leucas*) are able to transverse the barrier (as their size allows for a greater capacity to travel long distances). In support of this concept is the observation that, "endemic species are on average smaller than non-endemics".

2014

Structure and Dynamics of the Shark Assemblage off Recife, Northeastern Brazil
PLoS ONE, 9 (7): e102369
AFONSO, A.S. & ANDRADE, H.A. & HAZIN, F.H.V.

Understanding the ecological factors that regulate elasmobranch abundance in nearshore waters is essential to effectively manage coastal ecosystems and promote conservation. However, little is known about elasmobranch populations in the western South Atlantic Ocean. An 8-year, standardized longline and drumline survey conducted in nearshore waters off Recife, northeastern Brazil, allowed us to describe the shark assemblage and to monitor abundance dynamics using zero-inflated generalized additive models. This region is mostly used by several carcharhinids and one ginglymostomid, but sphyrnids are also present. Blacknose sharks, *Carcharhinus acronotus*, were mostly mature individuals and declined in abundance throughout the survey, contrasting with nurse sharks, *Ginglymostoma cirratum*, which proliferated possibly due to this species being prohibited from all harvest since 2004 in this region. Tiger sharks, *Galeocerdo cuvier*, were mostly juveniles smaller than 200 cm and seem to use nearshore waters off Recife between January and September. No long-term trend in tiger shark abundance was discernible. Spatial distribution was similar in true coastal species (i.e. blacknose and nurse sharks) whereas tiger sharks were most abundant at the middle continential shelf. The sea surface temperature, tidal amplitude, wind direction, water turbidity, and pluviosity were all selected to predict shark abundance off Recife.

Though not rare in Recife, *Carcharhinus leucas* and *C. limbatus* were comparably infrequent; only 11 Bull Sharks (and 6 Blacktips) were captured for this study. With such low numbers, Afonso et al. could not make specific determinations for either species in regard to their coastal environmental preferences; the abundance of *C. acronotus*, *Ginglymostoma cirratum* and *Galeocerdo cuvier* conversely yielded much more robust analyses. However, Afonso et al. do observe that the largest individual Bull Sharks were all female, and that, "bull sharks had the largest mean TL."

Nearshore habitat is almost synonymous with *Carcharhinus leucas*; it may be thus prudent to examine Afono et al.'s succinct characterization of coastal environs (as a way to better understand the Bull Shark's relationship to such complex, challenging but rewarding habitats). In comparison with other parts of the sea, nearshore areas are often highly productive and host to a dazzlingly array of fish and invertebrate biodiversity. For sharks, such habitats serve as ideal foraging grounds (especially in consideration of the fact that the coastal prey availability is likely far richer than that of the open ocean). These factors (in combination with their shallow depth) make coastal environments ideal pupping and nursery grounds for multiple nearshore species like *Carcharhinus leucas*.

Despite these considerable benefits, coastal habitats face two principal challenges: humans and dynamism. Shoreline human development has become a plague for coastal shark species, as it forces such nearshore denizens to cope with dams, canals, pollution, overfishing and damaged (or otherwise altered) habitat. These impacts greatly compound the inherent challenge of coastal dynamism—that is, the unstable fluctuation of nearshore environs as a result of tides, storms and seasonal variations in temperature. In sum, the coastal realm is a complex of riches and change—and to this, the Bull Shark is a mirror.

2014

Captura de Tiburones en la Región Noroccidental de Cuba
[Shark Catches in the Northwest Region of Cuba]
Latin American Jrnl. of Aquatic Research, 42 (3): 477-487

AGUILAR, C. & GONZALEZ-SANSON, G. & HUETER, R. &
ROJAS, E. & CABRERA, Y. & BRIONES, A. & BORROTO, R.
& HERNANDEZ, A. & BAKER, P.

Sharks have been important as seafood source and fisheries revenue in Cuba. Nevertheless, current information about this group of fishes in Cuba is scarce and in the last decades they have not been the focus of any organized research. From October 2009 to June 2011, fisheries and biological (229 sharks examined) data were collected at four landing sites in the northwest of Cuba. At present, there is no organized fishery specifically targeting only sharks along the northwest coast of Cuba, but they are caught as a component of multispecies fisheries on the insular shelf and as bycatch in longline fisheries targeting billfishes. We registered a total of 17 species, six in the commercial fishery, dominated by *Carcharhinus perezii*, *Sphyrna mokarran*, and *Carcharhinus leucas*, and 14 in the sport fishery (*i.e.*, small-scale artisanal, not recreational properly), dominated by *Isurus oxyrinchus*, *Isurus paucus*, *Carcharhinus longimanus*, *Carcharhinus falciformis*, *Galeocerdo cuvier* and *Prionace glauca*. Mean CPUE by months in sport fishing varied from 0.43 to 4.44 number of sharks caught per ten fishing trips. Most oceanic sharks caught in the Cuban sport fisheries are highly migratory species and their populations show great ecological connectivity throughout the Gulf of Mexico and adjacent waters.
Keywords: shark catches, species composition, Gulf of Mexico, Cuba.

Aguilar et al. have surprisingly little to say about *Carcharhinus leucas*, even though the species predominated within their recorded commercial catch (altogether, *C. leucas*, *C. perezii* and *Sphyrna mokarran* comprised 80% of the total commercial landings). Then again, only 20 individual Bull Sharks (known locally as "Cabeza de batea") were captured for this study: the commercial fishery landed 19 sharks (13 females and 6 males) while the recreational catch only yielded 1 male individual.

Bull Sharks aside, Anguilar et al. make fascinating determinations in regards to the other Cuban species and their relationships with the greater Gulf of Mexico and Caribbean community. While Caribbean Reef Sharks (*Carcharhinus perezii*) dominated the Cuban catch, Atlantic Sharpnose (*Rhizoprionodon terraenovae*) and Bonnethead (*Sphyrna tiburo*) Sharks were completely absent; this result is rather shocking, as Atlantic Sharpnose and Bonnethead Sharks are the Gulf of Mexico's most abundant species (while Caribbean Reef Sharks are by far less common). Anguilar et al. attribute this to Northwestern Cuba's unique ecology: the region is famous for its highly developed coral reefs—a habitat preferred by *Carcharhinus perezii*. Atlantic Sharpnose and Bonnethead Sharks conversely demonstrate a preference for estuarine areas and muddy bottoms; these two abundant species may thus be averse to visiting the Cuban reefs.

One of this study's most important findings was the observation of immature Oceanic Whitetip (*Carcharhinus longimanus*) and Silky (*Carcharhinus falciformis*) Sharks in Northwestern Cuba. Both species are coastal-pelagic, wide-ranging and severely overfished in the Gulf of Mexico: the presence of juveniles could thus be indicative of critical management habitat. Further investigation into Northwestern Cuba is essential to determine its potential role in the regional conservation of both species.

2014

Checklist of Chondrichthyans in Indian Waters
**Journal of the Marine Biological Association of India, 56
(1): 109-120**

AKHILESH, K.V. & BINEESH, K.K. & GOPALAKRISHNAN,
A. & JENA, J.K. & BASHEER, V.S. & PILLAI, N.G.K.

Conservation, management ans sustainable utilization of biological resources depend on the accurate identification of exploited taxa, which emphasizes the need for systematic taxonomic research. Chondrichthyans (sharks, rays, skates and chimaeras) are considered to be one of the most vulnerable exploited marine resources, however, the basic taxonomix study of these groups in Indian waters needs improvement to achieve better management for their sustainable exploitation. We discuss issues concerning chondrichthyan taxonomic research in India and provide an extended, updated checklist of chondrichthyans listed/reported from Indian waters, together with comments on their occurrence.

Keywords: Chondrichthyans, checklist, taxonomy, status, India, diversity, management, conservation.

330

Jena et al. reiterate the disturbing concerns of Alain Dubois by quoting, "this century has been called the century of extinctions." While no shark species has vanished from existence as of yet, many now endure the brink of collapse: the IUCN currently lists 17 species as Endangered (EN) and 10 species as Critically Endangered (CR) globally. Eight of these 27 (total) species are included within Jena et al.'s checklist for Indian seas:

Species	Validity in India	IUCN status (Global)
Holohalaelurus punctatus	Needs confirmation	EN
Carcharhinus hemiodon		CR
Glyphis gangeticus		CR
Glyphis glyphis	Needs confirmation	EN
Lamiopsis temminckii		EN
Sphyrna lewini		EN
Sphyrna mokarran		EN
Squatina squatina	Questionable	CR

At this critical point in global natural history, it is essential that scientists and conservationists fully understand shark biodiversity: as stated by the authors, "there is an urgent need for cataloguing biodiversity before several species become extinct without humans even knowing of their existence." *Carcharhinus leucas* may be relatively safe for now (recall that the IUCN considers the Bull Shark to be Near Threatened), but what of its derivatives? The concept of *Carcharhinus leucas* as a species complex is extremely recent, and the potential behind this idea is yet unresolved. If *Carcharhinus* cf. *leucas* 1 and *Carcharhinus* cf. *leucas* 2 are indeed separate species, would they too qualify as IUCN Near Threatened?

2014

Diversity of Trypanorhynch Metacestodes in Teleost Fishes from Coral Reefs off Eastern Australia and New Caledonia

Parasite (Paris, France), 21: 60

BEVERIDGE, I. & BRAY, R.A. & CRIBB, T.H. & JUSTINE, J.-L.

Trypanorhynch metacestodes were examined from teleosts from coral reefs in eastern Australia and from New Caledonia. From over 12,000 fishes examined, 33 named species of trypanorhynchs were recovered as well as three species of tentacularioids which are described but not named. Host-parasite and parasite-host lists are provided, including more than 100 new host records. Lacistorhynchoid and tentacularioid taxa predominated with fewer otobothrioid and gymnorhynchoids. Five species, *Callitetrarhynchus gracilis*, *Floriceps minacanthus*, *Pseudotobothrium dipsacum*, *Pseudolacistorhynchus heroniensis* and *Ps. shipleyi*, were particularly common and exhibited low host specificity. Limited data suggested a higher diversity of larval trypanorhynchs in larger piscivorous fish families. Several fish families surveyed extensively (Blenniidae, Chaetodontidae, Gobiidae, Kyphosidae and Scaridae) yielded no trypanorhynch larvae. The overall similarity between the fauna of the Great Barrier Reef and New Caledonia was 45%. Where available, information on the adult stages in elasmobranchs has been included.

Key words: Trypanorhyncha, Metacestodes, Great Barrier Reef, New Caledonia, Teleosts.

The Trypanorhyncha are an order of parasitic flatworms that commonly infect coral reef fishes as larvae and later parasitize elasmobranchs as adults. Larval Trypanorhynchs occur, "most commonly in the body cavity but may also be found in the musculature or other sites such as the gill arches." Once consumed by sharks and other elasmobranchs (via their infected host), Trypanorhynchs transition into the adult form and subsequently parasitize the consumers' stomach or spiral valves.

Two adult Trypanorhynchs infect *Carcharhinus leucas*. The cosmopolitan *Callitetrarhynchus gracilis* demonstrated low specificity but nevertheless, "exhibited the widest host range." Intermediate stages of this species were recorded in approximately 130 teleosts, spanning 18 families and 5 orders; adult *C. gracilis* utilize requiem sharks (Carcharhinidae) as the, "primary definitive hosts in the Australian region." The more locally distributed *Floriceps minacanthus* (limited to the Indo-Pacific region) was recorded from a much smaller number of teleosts spanning only 6 families and 2 orders. In previous accounts, adult *F. minacanthus* have only been reported from four species of *Carcharhinus*; however, Beveridge et al. report *F. minacanthus* from *Triaenodon obesus* (Whitetip Reef Shark)—the first non-*Carcharhinus* host recorded in the literature.

As Trypanorhynchs are rather difficult to identify, they have oftentimes been overlooked as a component of invertebrate biodiversity. However, in light of mounting threats to coral reef health and ecology (especially so in the case of the Great Barrier Reef), there has been a renewed interest for, "focusing attention on the full diversity of reefs rather than simply on the diversity of fish and corals, the most obvious examples of reef diversity." Vertebrate endoparasites are part of a "hidden" biodiversity—a set of secrets trapped within what is seen...what benefits could come of their illumination?

2014

Landing Trends, Species Composition and Percentage Composition of Sharks and Rays in Chittagong and Cox's Bazar, Bangladesh

International Journal of Advanced Research in Biological Sciences, 1 (3): 81-93

BIKRAM, R. & ALAM, F. & RHAMAN, G. & SINGHA, N.K. & AKHTAR, A.

The study was conducted from April, 2006 to June, 2010 on landing trends, species composition and percent contribution of sharks and rays by weight using the catch records of Marine Fisheries Survey Management Unit, Chittagong from two landing centers Fishery ghat, Chittagong and BFDC ghat, Cox's Bazar. This study identified 27 species in total representing 11 species of shark (04 families) and 16 species of ray (09 families). The highest landing volume (134 MT) and contribution (76%) to total catch for the whole sampling period was found from *Scoliodon laticaudus* followed by *Rhizoprionodon acutus* (108 MT or 55%), *Carcharhinus melanopterus* (75 MT or 38%), *Sphyrna zygaena* (49 MT or 26%), *Chiloscyllium indicum* (38 MT or 20%), *Eusphyra blochii* (22 MT or 11%) *Galeocerdo cuvier* (21 MT or 10%) and other (03 MT or 2%). Species which occurred least were placed in the 'other' category comprising *Carcharhinus amblyrhynchos*, *Stegostoma fasciatum*, *Carcharhinus leucas* and *C. falcifomis*.

Keywords: Shark, ray, species composition, elasmobranches fishery, landing tend, catch data, abundance, percent contribution.

334

According to Bikram et al., Bangladesh is in urgent need of a National Plan of Action (NPOA) for the sustainable development and management of its local shark fisheries. While sharks and rays only comprise 1% of the total Bangladeshi marine landings, this number equates to a mighty catch (whose dynamics are not fully understood as a result of inconsistent recordkeeping and confusion in regards to species identification). Indeed, Bikram et al. specifically request for the improvement of local taxonomic training and the development of descriptive species profiles (ideally in the form of a field guide) complete with, "local names, valid scientific names, pictures and status in IUCN red list." Such improvements (as perhaps a regional FAO field guide could provide) are of pressing demand, as the authors are currently unable to extrapolate their data to sufficiently assess the population trends of Bangladeshi sharks (though it is mentioned that the, "total landing of sharks and rays were found to be on slight decline.")

Carcharhinus leucas is a rare component of the Bangladeshi catch, and only appears in the July 2009-June 2010 portion of the authors' dataset (comprising only 0.31% of the elasmobranch landings for that timespan—a total catch of 0.527 MT). Bikram et al. suggest that the relative absence of large sharks (greater than 100kg) from Bangladeshi landings may be due to the fact that such specimens are, "mostly common in offshore water which is beyond the reach of our artisanal fishermen". Most of the catch from this study was indeed acquired from artisanal fisheries utilizing gill nets, hooks and lines, trammel nets and set bag nets within the 40-meter depth contour. However, as *Carcharhinus leucas* is a primarily coastal species, it may simply be less abundant in Bangladeshi waters by nature (especially in consideration of the fact that shark fishing is conducted throughout the year, with the main season lasting from November to March).

2014

Long-Term Changes in Species Composition and Relative Abundances of Sharks at a Provisioning Site
PLoS ONE, 9 (1): e86682
BRUNNSCHWEILER, J.M. & ABRANTES, K.G. & BARNETT, A.

Diving with sharks, often in combination with food baiting/provisioning, has become an important product of today's recreational dive industry. Whereas the effects baiting/provisioning has on the behaviour and abundance of individual shark species are starting to become known, there is an almost complete lack of equivalent data from multi-species shark diving sites. In this study, changes in species composition and relative abundances were determined at the Shark Reef Marine Reserve, a multi-species shark feeding site in Fiji. Using direct observation sampling methods, eight species of sharks (bull shark *Carcharhinus leucas*, grey reef shark *Carcharhinus amblyrhynchos*, whitetip reef shark *Triaenodon obesus*, blacktip reef shark *Carcharhinus melanopterus*, tawny nurse shark *Nebrius ferrugineus*, silvertip shark *C. albimarginatus*, sicklefin lemon shark *Negaprion acutidens*, and tiger shark *Galeocerdo cuvier*) displayed inter-annual site fidelity between 2003 and 2012. Encounter rates and/or relative abundances of some species changed over time, overall resulting in more individuals (mostly *C. leucas*) of fewer species being encountered on average on shark feeding dives at the end of the study period. Differences in shark community composition between the years 2004-2006 and 2007-2012 were evident, mostly because *N. ferrugineus*, *C. albimarginatus* and *N. acutidens* were much more abundant in 2004-2006 and very rare in the period of 2007-2012.

Carcharhinus leucas is the Shark Reef Marine Reserve's most abundant species, and has steadily increased in number throughout the course of Brunnschweiler et al.'s eight-year study period (from 2004 to 2012). *Nebrius ferrugineus*, *Carcharhinus albimarginatus* and *Negaprion acutidens* have all simultaneously decreased in abundance at the Shark Reef Marine Reserve, but have conversely increased at the nearby Lake Reef (a competitor dive operator with a dissimilar feeding protocol). Intriguingly, the number of *C. leucas* at Lake Reef has apparently, "dropped over time".

Brunnschweiler et al. propose a fascinating hypothesis for these observations: *Carcharhinus leucas* may be competitively excluding the other three species from feeding at Shark Reef (and could be simultaneously diplacing them into the Lake Reef area). As a larger species, the Bull Shark may have secured itself at the peak of a local dominance hierarchy. In support of this idea is the fact that the smaller shark species at Shark Reef were, "never observed approaching the feeder at the 30 m and 16 m feeding sites where the larger species are fed." Indeed, both *C. albimarginatus* and *N. acutidens* have specifically decreased in relative abundance at the 16 m feeding site; anecdotally, it has also been observed among local divers that, "as numbers of *C. leucas* increased over the years, *C. albimarginatus* and *N. acutidens* were less able to approach the feeder."

The increase of *Carcharhinus leucas* at Shark Reef (as a result of regular feeding since 2003) has been an unquestionable benefit to local ecotourism. However, this newfound concentration of Bulls may be ecologically disruptive on two fronts. Firstly, the feeding attraction of Shark Reef may have 'siphoned' Bull Sharks from other areas (such as, quite possibly, Lake Reef). Secondly, this same attraction may be indirectly fueling the displacement of other shark species from Shark Reef (reducing the area's total biodiversity).

2014

Residency Patterns and Migration Dynamics of Adult Bull Sharks (Carcharhinus leucas) on the East Coast of Southern Africa

PLoS ONE, 9 (10): e109357

DALY, R. & SMALE, M.J. & COWLEY, P.D. & FRONEMAN, P.W.

Bull sharks (*Carcharhinus leucas*) are globally distributed top predators that play an important ecological role within coastal marine communities. However, little is known about the spatial and temporal scales of their habitat use and associated ecological role. In this study, we employed passive acoustic telemetry to investigate the residency patterns and migration dynamics of 18 adult bull sharks (195-283 cm total length) tagged in southern Mozambique for a period of between 10 and 22 months. The majority of sharks (n = 16) exhibited temporally and spatially variable residency patterns interspersed with migration events. Ten individuals undertook coastal migrations that ranged between 433 and 709 km (mean = 533) with eight of these sharks returning to the study site. During migration, individuals exhibited rates of movement between 2 and 59 km.d^{-1} (mean =17.58 km.d^{-1}) and were recorded travelling annual distances of between 450 and 3760 km (mean =1163 km). Migration towards lower latitudes primarily took place in austral spring and winter and there was a significant negative correlation between residency and mean monthly sea temperature at the study site. This suggested that seasonal change is the primary driver behind migration events but further investigation is required to assess how foraging and reproductive activity may influence residency patterns and migration.

Daly et al.'s investigation takes place within the Ponta do Ouro Partial Marine Reserve, and primarily focuses on *Carcharhinus leucas* from the Pinnacle Reef complex in southern Mozambique. Pinnacle Reef is a fossilized coastal dune consisting of, "a ridge approximately 1.2 km long with a series of shallower pinnacles along its spine". According to the authors, this area appears to be an important aggregation site for adult *Carcharhinus leucas* as the sharks, "exhibit prolonged periods of residency, primarily during the austral summer months." While there is a high amount of variance between individual preferences, most of the Pinnacle Bulls appeared to favor the reef during daylight hours. This temporal use of habitat is most likely driven by foraging behavior (rather than predator avoidance, as the local Bull Sharks lack predator constraints).

Trevally (specifically *Caranx ignobilis*, *Carangoides fulvoguttatus*, *Carangoides gymnostethus* and *Caranx sexfasciatus*) may be (in part) compelling this foraging behavior as, "the onset of increasing levels of residency exhibited by bull sharks at the study site appeared to be associated with large aggregations of Carangid fish species". However, Daly et al. additionally claim that, "it was not clear if bull sharks were responding to increased prey abundance or if both the sharks and the fish aggregations were corresponding to environmental cues." Both predator and prey may be responding to seasonal changes in temperature; the abundance of *Carcharhinus leucas* at Pinnacle Reef peaks during austral summer, but soon after declines as temperatures cool. In consistence with prior studies, Daly et al. affirm that the majority of Pinnacle Bulls undertake a, "northward migration to warmer latitudes during austral winter and spring". Interestingly, female sharks were found to travel greater distances than males, and were furthermore recorded "undertaking the majority of migration events (70%)".

2014

Quantifying Shark Distribution Patterns and Species-Habitat Associations: Implications of Marine Park Zoning
PLoS ONE, 9 (9): e106885
ESPINOZA, M. & CAPPO, M. & HEUPEL, M.R. & TOBIN,
A.J. & SIMPFENDORFER, C.A.

Quantifying shark distribution patterns and species-specific habitat associations in response to geographic and environmental drivers is critical to assessing risk of exposure to fishing, habitat degradation, and the effects of climate change. The present study examined shark distribution patterns, species-habitat associations, and marine reserve use with baited remote underwater video stations (BRUVS) along the entire Great Barrier Reef Marine Park (GBRMP) over a ten year period. Overall, 21 species of sharks from five families and two orders were recorded. Grey reef *Carcharhinus amblyrhynchos*, silvertip *C. albimarginatus*, tiger *Galeocerdo cuvier*, and sliteye *Loxodon macrorhinus* sharks were the most abundant species (>64% of shark abundances). Multivariate regression trees showed that hard coral cover produced the primary split separating shark assemblages. Four indicator species had consistently higher abundances and contributed to explaining most of the differences in shark assemblages: *C. amblyrhynchos*, *C. albimarginatus*, *G. cuvier*, and whitetip reef *Triaenodon obesus* sharks. Relative distance along the GBRMP had the greatest influence on shark occurrence and species richness, which increased at both ends of the sampling range (southern and northern sites) relative to intermediate latitudes. Hard coral cover and distance across the shelf were also important predictors of shark distribution.

Only four adult *Carcharhinus leucas* were recorded by Espinoza et al.'s BRUVS array (comprising only 0.5% of the total observed shark biodiversity). The Bull Shark thus does not appear to be a major component of the Great Barrier Reef Marine Park community. The four individuals in this study were observed from a wide variety of habitats (including inshore areas, shelf areas and the coral reef itself) at depths between 17.7 m and 34.7 m. All four Bull Sharks were recorded from open fishing areas. Like *G. cuvier* and *Sphyrna mokarran*, *C. leucas* is exemplified by the authors as a more mobile (and less reef-bound) species with a broad habitat range.

Overall, sharks were generally rare or uncommon in the GBRMP and occurred in only 25% of the entire BRUVS array. Most BRUVS only recorded one or two species at a time, but some sites had, "disproportionally higher shark species richness." Structurally complex areas located offshore (such as rocky shoals, coral reefs and habitats dominated by marine plants) had, "more species of sharks than coastal inshore habitats with lower complexity". Espinoza et al. claim that such complex areas (in combination with strong current flow) are known for their "predictable aggregations" of both local and non-resident reef shark species (which, presumably, find this complex ideal for foraging).

Of utmost importance is the finding that shark abundances were, "significantly greater in areas closed to fishing, and the effect was significantly greater in sites with higher coral cover." This latter point is of particular concern as coral bleaching, crown-of-thorns starfish predation and tropical cyclones have together, "resulted in a 50% decline of coral cover within the GBR over the past two decades". The comprehensive protection of GBRMP sharks thus requires a combination of effective fishing regulation (such as no-take and no-entry zones) and careful management of the reefs themselves.

2014

Bull Shark (Carcharhinus leucas) Exclusion Properties of the Sharksafe Barrier and Behavioral Validation Using the ARIS Technology

Global Ecology and Conservation, 2: 300-314

O'CONNELL, C.P. & HYUN, S.-Y. & RILLAHAN, C.B. & HE, P.

Magnetic deterrents have recently been employed to assess their ability to reduce elasmobranch mortality in beach nets. With previous studies exhibiting promise, the present study examined the ability of a magnetic barrier technology, known as the Sharksafe Barrier, to exclude bull sharks (*Carcharhinus leucas*) from bait, and how behavioral interactions may change with variations in environmental and biological factors. Generalized linear mixed model analyses based on 114, 30-min trials illustrate that all interacting *C. leucas* were successfully excluded from baited procedural control and magnetic regions (i.e. zero entrances through either region). Avoidance and pass around frequencies significantly differed from the control region and were based on situational context. To enhance behavioral analysis techniques, an Adaptive Resolution Imaging Sonar (ARIS) was employed which revealed that *C. leucas* distance from and swim speed associated with the magnetic barrier region were significantly greater than those associated with the procedural control region. This study demonstrates the Sharksafe barrier's effectiveness in excluding *C. leucas* from baited regions, regardless of variations in biological and/or environmental parameters. While other bather protection systems (e.g. beach nets and drumlines) continue to be used, this study exhibits promise that the Sharksafe barrier can be an eco-friendly alternative to beach nets.

The ampullae of Lorenzini are keys to the Bull Shark's 6th sense: electroreception. Pores scattered across the Bull Shark's head serve as the entrances to the ampullae of Lorenzini (which are themselves small canals filled with a low-resistance hydrogel). When the Bull Shark encounters an electrical stimulus, its surface pores receive a voltage potential that differs from that of the inner ampullae canals; this difference, "elicits a neurological impulse that is sent to the brain where the stimulus is perceived." Thus, the Bull Shark is able to detect an amazing array of stimuli, ranging from the electric fields of predators, prey and other Bull Sharks to (presumably) the Earth's magnetic field.

Earth's geomagnetic flux naturally ranges from 0.25 to 0.65 gauss (G). O'Connell et al. brilliantly apply the comparatively enormous flux of permanent magnets (specifically, grade C8 barium ferrite magnets, which have a surface magnetic flux of 3,850 G) to "overstimulate" the Bull Shark's ampulatory system (and thereby repulse the animal). The Sharksafe Barrier (crafted from a combination of these permanent magnets, PVC piping and foam buoys) appears to succeed in this endeavor. For this particular study, O'Connell et al. designated three experimental areas: the magnetic area (to which the fully magnetized Sharksafe Barrier was applied); the procedural control (to which a visually identical but unmagnetized 'mock barrier' was applied); and the control (an untreated area). A total of 25 Bull Sharks were observed throughout the course of this study, and *none of them* entered either the magnetic barrier or procedural control areas (by fantastic contrast, these same 25 Bull Sharks made a total of 1,260 entries into the unprotected control area). These findings explode the collective hope for an ecologically friendly protective barrier between sharks and humans—a barrier that harms neither party and furthermore protects collateral targets (like marine mammals) from incidental catch.

2014

Elasmobranch Fishery Resources of Gulf of Mannar,
Southeast Coast of India
World Journal of Fish and Marine Sciences, 6 (1): 24-29
THEIVASIGAMANI, M. & SUBBIAH, S.

Various species of elasmobranchs are fishing out by fisher folk with increasing fishing pressure. The cartilaginous fishery resources were studied from April 2012 to October 2012 from Gulf of Mannar. The samples were identified based on their morphological characters and grouped under their family and orders. The rays shared first position and sharks contribute next among all the elasmobranch fishery resources of this region. There are 6 orders lamniformes, carcharhiniformes, orectolobiformes, myliobatiformes, pristiformes and rajiformes that hold 65 species of various elasmobranchs landed during this research. Myliobatiformes and carcharhiniformes represents 49.23% and 32.31% respectively, regarding species diversity is concerned.

Key words: Gulf of Mannar • Elasmobranchs • Carcharhinidae • Dasyatidae

344

The Gulf of Mannar's Exclusive Economic Zone (EEZ) spans to about 15,000 km^2; of this, the Gulf of Mannar Biosphere Reserve (the first such reserve in Southeast Asia) protects 10,500 km^2 (a vast majority of the EEZ). Under the guidance of the UNESCO's Man and Biosphere Program, the Indian government established the Gulf of Mannar Biosphere Reserve in 1989. The Reserve spans, "from Rameswaram to Tuticorin, lies between 78°5'E-79°30'E and 8°45'N-9°25'N and extends to a distance of 140 km." Twenty-one islands rimmed with maritime forests, beaches, mudflats, salt marshes, mangroves, estuaries, sea grasses and coral reefs characterize the rich biodiversity of this region, and furthermore protect the Gulf coastline from serious storm action. Likewise, the islands and coral reefs, "offer shelter for a variety of marine fauna and flora".

Carcharhinus leucas is one such faunal element of the Gulf of Mannar Biosphere Reserve. Theivasigamani & Subbiah offer no specific details in regards to this species' wellbeing, population dynamics or ecological impact; instead, they report on the general presence of 26 total shark species landed by local fishermen. While the Biosphere itself is relatively protected, the surrounding reef and nearshore communities (spanning a total of 5,500 km^2) are vulnerable to overfishing. Gulf fisheries are conducted year round by both mechanized and nonmechanized vessels utilizing a mix of trawl nets (32% of shark landings) gill nets (31%) and hooks and lines (20%). Shore seining exerts a seasonal impact (particularly in the southern portions of the Gulf) and additionally contributes to the area's mounting fishing pressure. Theivasigamani & Subbiah cite a global shift towards the commercial interest in sharks (particularly for their fins and oil) as a primary impetus for their increased overexploitation in the Gulf. The Whale Shark (*Rhincodon typus*) elicits particular concern, as the species has become, "an easy target since 1980".

2015

At-Vessel Mortality and Blood Biochemical Status of Elasmobranchs Caught in an Australian Commercial Longline Fishery

Global Ecology and Conservation, 3: 878-889
BUTCHER, P.A. & PEDDEMORS, V.M. & MANDELMAN, J.W. & MCGRATH, S.P. & CULLIS, B.R.

This study investigates mortality of sharks in a commercial longline fishery in Australia. To examine the rate and biological, environmental and technological factors contributing to at-vessel mortality, four setlines with 120 gangions possessing 'hook timers' were deployed daily using conventional gears from two commercial fishing vessels during 2013. A total of 689 animals across 22 species and including 18 elasmobranchs were landed. For the five species (*Carcharhinus* spp.), and one genus (*Sphyrna* spp) where there were sufficient numbers for analysis, generalized linear mixed models showed that species and the elapsed time spent on the line after hooking were the strongest predictors of at-vessel mortality, with spinner (*C. brevipinna*), blacktip (*C. limbatus*) and hammerhead (*Sphyrna* spp) sharks exhibiting the highest death rates. The variables which best explained mortality, included: (i) sex of the caught sharks, and the interaction between species with (ii) capture depth, and (iii) the elapsed time spent on the line after hooking. For the subset of dusky (*C. obscurus*) and sandbar (*C. plumbeus*) sharks examined for physiological status at the point of capture, very few of the 13 chosen blood analytes varied significantly. Given the observed high mortality rates and stress associated with the time spent on the line after capture, operational changes to reduce these adverse impacts should be considered.

346

Previous capture mortality studies have suggested that *C. leucas* is a relatively hardy species able to endure prolonged soak times. While only one Bull Shark was captured within this particular study, its health appeared to be in characteristically good order: even after an extensive soak time of 14 hours and 10 minutes, the 311 cm Bull Shark still remained in high condition (with no observable wounds, bleeding, skin damage or bruising) and furthermore displayed a moderate level of activity upon release. This behavior is much in contrast to that of Sandbar, Dusky, Spinner, Blacktip (here referred to as Common Blacktip) and Hammerhead (Great, Smooth and Scalloped) Sharks; death and high capture stress were recorded for each of these species (which are, uncoincidentally, obligate ram ventilators). Unlike Tiger, Grey Nurse (Sandtiger), Wobbegong and Bull Sharks (which each demonstrated low capture stress and high survival rates), these species are incapable of buccal pumping and are thus required to swim continuously in order to push water over their buccal cavities and gills.

The spatial restrictiveness and resulting damage that longlines deal to obligate ram ventilators is particularly alarming when it comes to the three Hammerhead species: the Great and Scalloped are listed as Endangered species (IUCN) while the Smooth listed as Vulnerable. Butcher et al. propose that the simplest and most 'shark-minded' change to longline fishing would be to, "shorten the duration of line deployment (i.e. soak time)." By limiting deployment times to 2 hours (conservatively) or 5 hours (realistically), highly sensitive species like the Hammerheads would endure less soak time capture stress and thus likely benefit from increased survivorship. Butcher et al. additionally suggest temporal closures for Hammerhead areas, stating that it would be, "necessary to reduce their capture all together by fishing at night when they are less likely to be hooked."

2015

Sawfishes and Other Elasmobranch Assemblages from the
Mio-Pliocene of the South Caribbean
(Urumaco Sequence, Northwestern Venezuela)
PLoS ONE, 10 (10): e0139230
CARRILLO-BRICEÑO, J.D. & MAXWELL, E. & AGUILERA,
O.A. & SÁNCHEZ, R. & SÁNCHEZ-VILLAGRA, M.R.

The Urumaco stratigraphic sequence, western Venezuela, preserves
a variety of paleoenvironments that include terrestrial, riverine, lacustrine
and marine facies. A wide range of fossil vertebrates associated with these
facies supports the hypothesis of an estuary in that geographic area
connected with a hydrographic system that flowed from western Amazonia
up to the Proto-Caribbean Sea during the Miocene. Here the elasmobranch
assemblages of the middle Miocene to middle Pliocene section of the
Urumaco sequence (Socorro, Urumaco and Codore formations) are
described. Based on new findings, we document at least 21 taxa of the
Lamniformes, Carcharhiniformes, Myliobatiformes and Rajiformes, and
describe a new carcharhiniform species († *Carcharhinus caquetius* sp.
nov.). Moreover, the Urumaco Formation has a high number of well-
preserved fossil *Pristis* rostra, for which we provide a detailed taxonomic
revision, and referral in the context of the global Miocene record of *Pristis*
as well as extant species. Using the habitat preference of the living
representatives, we hypothesize that the fossil chondrichthyan assemblages
from the Urumaco sequence are evident for marine shallow waters and
estuarine habitats.

A total of twenty-eight *Carcharhinus leucas* teeth were recovered from both the Urumaco (middle-late Miocene) and Codore (late Miocene-Pliocene) formations in western Venezuela (specifically from the Urumaco town area, in the state of Falcón). The presence of this species hints of an ancient estuary or nearshore habitat; in correlation, Carrillo-Briceño et al. additionally uncover *Galeocerdo cuvier*, *C. limbatus*, *C. porosus* and *Negaprion brevirsotris* teeth from the Urumaco sequence, as well as the coastal Batoid dentitions of *Aetobatus narinari*, *Rhinoptera*, *Myliobatis*, *Dasyatis* and *Pristis*. This characteristically nearshore Elasmobranch assemblage is complimented with the contemporaneous fossils of other coastal taxa such as sea cows, turtles, crocodiles and bony fish characteristic of lacustrine, riverine, estuarine and shallow marine habitats.

In respect to both this sweeping fossil assemblage and the paleoenvironment as characterized by previous studies, Carrillo-Briceño et al. hypothesize that, "the fossil elasmobranch assemblages from the Urumaco sequence are associated primarily with shallow marine and estuarine environments". This hypothesis potentially offers new insight into the paleoecology of truly extinct Elasmobranch lineages (lacking a modern counterpart); the presence of *Carcharocles megalodon*, *Hemipristis serra* and (the newly described) *Carcharhinus caquetius* sp. nov. in Urumaco may indicate that these species are likewise characteristic pillars of the coastal (and potentially estuarine) communities of the Venezuelan Miocene and Pliocene. While large marine cetaceans—the most charismatic prey of *Carcharocles megalodon*—are relatively absent from the Urumaco sequence, the presence of, "*C. megalodon* in the same strata as bony fishes, turtles, sirenids, and crocodiles suggests that these could possibly have been prey for this shark." Juvenile sirenids may have likewise been potential prey items for *Carcharhinus leucas*.

2015

Global Pattern of Phylogenetic Species Composition of
Shark and Its Conservation Priority
Ecology and Evolution, 5 (19): 4455-4465
CHEN, H. & KISHINO, H.

The diversity of marine communities is in striking contrast with the diversity of terrestrial communities. In all oceans, species richness is low in tropical areas and high at latitudes between 20 and 40°. While species richness is a primary metric used in conservation and management strategies, it is important to take into account the complex phylogenetic patterns of species compositions within communities. We measured the phylogenetic skew and diversity of shark communities throughout the world. We found that shark communities in tropical seas were highly phylogenetically skewed, whereas temperate sea communities had phylogenetically diversified species compositions. Interestingly, although geographically distant from one another, tropical sea communities were all highly skewed toward requiem sharks (Carcharhinidae), hammerhead sharks (Sphyrnidae), and whale sharks (*Rhincodon typus*). Worldwide, the greatest phylogenetic evenness in terms of clades was found in the North Sea and coastal regions of countries in temperate zones, such as the United Kingdom, Ireland, southern Australia, and Chile. This study is the first to examine patterns of phylogenetic diversity of shark communities on a global scale. Our findings suggest that when establishing conservation activities, it is important to take full account of phylogenetic patterns of species composition and not solely use species richness as a target.

In 2011, the Global Shark Distribution Database (Lucifora et al.)[11] superimposed the distributions of 507 shark species (the entirety of known shark biodiversity) onto a single map. This effort was unprecedented in its graphical representation of species richness (the primary index of biodiversity), and revealed shark biodiversity hotspots in the temperate waters of Taiwan, Japan and eastern Australia. Chen & Kishino build upon this endeavor with a twofold global masterpiece: a complete phylogenetic assessment of 236 shark species (based on their COI mtDNA sequences) and a subsequent application of this assessment to multiple indices of global biodiversity (complete with maps charting the superimposed distributions of community richness, distinctiveness and composition). This complex analysis involves four phylogenetic diversity indices:

PD	*Phylogenetic Diversity*	Community phylogenetic richness
AvTD	*Avg. Taxonomic Distinctiveness*	Community phylogenetic distinctiveness
PS	*Phylogenetic Skew*	Community species composition (skewed)
PE	*Phylogenetic Evenness*	Community species composition (even)

High PS values are indicative of a phylogentically "skewed" community (that is, a community phylogenetically biased towards a particular taxon like the Carcharhinidae) while low PS values are indicative of a phylogentically "balanced" or "even" community (incorporating shark species across multiple families and orders). PE is the inverse of PS, and was highest in temperate waters, while PS was highest along tropical coasts. PD was greatest in eastern Australia, while AvTD was highest on the east coast of Japan. The variance in these findings illustrates the need to, "consider other biodiversity hotspots". Proper conservation must be considerate of these indices (and not only species richness).

2015

Risk Perceptions and Conservation Ethics among
Recreational Anglers Targeting Threatened Sharks in the
Subtropical Atlantic
Endangered Species Research, 29 (1): 81-93
GALLAGHER, A.J. & COOKE, S.J. & HAMMERSCHLAG, N.

Recreational fisheries management has traditionally been more concerned with quantifiable, catch-centric goals than angler-centric perceptions. However, the attitudes of fishers affect their behavior, which can alter the effort they make towards conservation actions, and ultimately, the outcome for exploited or threatened species. We conducted a quantitative human dimensions study into the drivers of conservation attitudes and perceptions of recreational fishers towards sharks. This was accomplished through a targeted online snowball survey on a sample of 158 recreational anglers in the state of Florida, a global hotspot for recreational fishing. Subjective knowledge of shark conservation issues was the most consistent driver for pro-shark conservation attitudes. Anglers ranked the great hammerhead and tiger shark as being the most threatened species, a result that is generally consistent with empirical data. Anglers did not identify species-specific differences in capture stress as an important factor in determining survivability, a result that somewhat contradicts available empirical data. In general, fishers were more supportive of management actions that would be the least restrictive to fishing, except in the case of highly threatened species. Anglers believed commercial fishing had the largest impact on shark populations, and recreational fishing the least, which is largely consistent with empirical information but could also reflect angler bias.

The angler's perspective on shark fishing carries a striking amount of power. In the most direct sense, an angler's underlying beliefs and attitudes form a pattern of behavior that determines the fate of his catch: life and death are literally in the balance. As explained by Gallagher et al., this multifaceted concept is, "best exemplified when anglers choose to remove their catch from the water for photos, a conscious decision that puts the fate of the fish in the hands of the angler prior to release." Notoriously sensitive species like *Sphyrna mokarran* (Great Hammerhead) would not be able to cope well with such a physiological and metabolic stressor; conversely, hardy species like *Carcharhinus leucas* may better endure an out-of-water photo opportunity (though the event would nonetheless be stressful). Species-specific preferences like these go largely unnoticed within the recreational shark fishing community (even amongst experienced anglers).

While a lack of awareness of these nuances is of particular concern (especially in light of the fact that the Great Hammerhead is endangered, while many other Florida species face some form of threat), Gallagher et al. note that the anglers sampled in their survey were, "relatively well-informed on issues pertaining to shark conservation, although they believed their actions to have minimal impacts on shark populations." Indeed, this minimization of angler impact is prevalent, and may stem from two chief motivations: a sincere avoidance of blame, and/or a, "perceived fear of incurring fishing restrictions." Restrictive (but ecologically effective) conservation measures were largely unpopular within the survey sample; the superlative conservation effort that is the marine protected area was, "the least popular management tool, a finding which agrees with the controversy surrounding marine protected areas in recreational fisheries." Thus is the compelling (and complicated) crux: how do we preserve our natural resources while protecting our personal freedoms?

2015

Sharks of the Arabian Seas: an Identification Guide
The International Fund for Animal Welfare,
Dubai, UAE 240 pp.
JABADO, R.W. & EBERT, D.A.

BULL SHARK *Carcharhinus leucas*   (Müller & Henle 1839)

FAO code: **CCE**

NT (2005)

KEY FEATURES

1 Snout broad, short and blunt, its length less than mouth width

2 First dorsal fin high and triangular, height equal or less than 3.1 times height of second dorsal fin; its origin over inner margins of pectoral fins

3 Second dorsal fin with concave upper margin and short rear tips, its origin anterior to
  anal fin origin

4 Pectoral fins large and angular with straight edges

5 No interdorsal ridge

6 Juveniles with dusky fin tips, less prominent in adults

SIZE

B: 55—81 cm. **Mat:** ♀180—230 cm, ♂157—226 cm. **Max TL:** 340 cm.

HABITAT

Found inshore in turbid or brackish waters, hypersaline lagoons, bays and canals. Occurs near the bottom from the surf line to a depth of at least 152 m.

Carcharhinus leucas of the Arabian Seas Region can be found in the Arabian Sea proper (including the span from northern Somalia to Ras al Hadd in Oman, but excluding the Gulf of Aden), the Arabian (Persian) Gulf, the Strait of Hormuz, the Gulf of Oman, and the Gulf of Kutch. This species is reported absent from the Red Sea, and (in such case) would likely be isolated from the area as a result of its semi-enclosed biogeographic nature. It should be noted, however, that as research on sharks of the Arabian seas is limited, many species could have a, "wider distribution than shown here." Indeed, while the absence of *Carcharhinus leucas* from the Red Sea is corroborated by the IUCN's official rangemap, it remains unsupported by the FAO (which affirms the that Bull Shark is present throughout the majority of the Red Sea) [12]. In entertainment of this possibility, Jabado & Ebert state that, "gaps in species distribution, or patchy distributions, does not mean that they do not occur in other areas of the Arabian Seas but rather that we may not have enough knowledge to accurately assess the occurrence of these species."

Guides such as *Sharks of the Arabian Seas* are thus critical to facilitate the efforts of scientists and conservationists to accurately characterize a region's biodiversity and community composition (primarily in the pursuit of responsible resource management). While most fisheries of the Arabian Sea are artisanal, there is nonetheless demonstrable evidence of regional overexploitation. Four international agreements currently guide the conservation of sharks from the Arabian seas: the Convention on the Conservation of Migratory Species of Wild Animals (CMS), The Memorandum of Understanding (MOU) on the Conservation of Migratory Sharks, The Convention on International Trade in Endangered Species of Wild Fauna and Flora (CITES), and the Indian Ocean Tuna Commission (IOTC). None of these agreements currently apply to *Carcharhinus leucas*.

2015

Guidance on National Plan of Action for Sharks in India
CMFRI Marine Fisheries Policy Series (2): 1-102
KIZHAKUDAN, S.J. & ZACHARIA, P.U. & THOMAS, S. &
VIVEKANANDAN, E. & MUKTHA, M.

India is one of the major shark fishing nations of the world, contributing to about 9% of the global catch of sharks during 2000-2009 with an average annual production of 54,614 t. Sustainable shark fishing was practised in India by artisanal fishermen before the introduction of mechanised fishing, which led to sharks being landed as by-catch. Later, in the 1990s, targeted shark fishing began when the demand for sharks increased in international markets. Although there was increase in shark catches initially there has been a consistent decline in the last one decade which has raised serious concern on this resource.

In 2001, India joined other nations in conserving sharks by protecting ten species under Schedule I of the Indian Wildlife (Protection) Act, 1972. India is also a signatory party to the recent CITES Appendix II listing of 5 species of sharks (of which 4 species are commonly found in Indian waters) and 2 species of manta rays, thereby initiating regulation of fin and gill plate trade in these species. Shark finning and export/import of shark fins are also prohibited in India. However, strategies to avoid protected or trade-regulated species from capture in directed as well as multispecies fisheries do not exist.

The Central Marine Fisheries Research Institute has served as a pioneering research institute in India working on fishery dependent daya analysis for resource assessment of sharks along the Indian coast.

Carcharhinus leucas is one of 88 shark species reported from Indian waters by the Kochi-based Central Marine Fisheries Research Institute (CMFRI). No level of international or local protection is offered for this species, and while Kizhakudan et al. present little detail on the Bull Shark's specific fisheries dynamics, they do note that the genus *Carcharhinus* dominates the shark landings of India's southern coasts, and that the requiem sharks (Carcharhinidae) are usually deemed to possess second-class shark fins. Nevertheless, *Carcharhinus leucas* is a common catch in Indian fisheries, and as such, Kizhakudan et al. do provide information on some of the Bull Shark's biological indices:

Length range in fishery	224-327 cm
Mean length in fishery	260 cm
Maximum length (Lmax)	360 cm
Lm50	193 cm
Number of young ones	1-13
Gestation period	10-11 months
Estimated Longevity	32 years

India is currently home to the world's second largest shark fishery (with Indonesia being largest). While the Indian export of shark fins in 2011 was about 195 t (valued at roughly US $15 million), the country itself does not have a domestic fin market, and has recently prohibited the export and import of shark fins in India (by order of the Government of India's Department of Commerce of the Ministry of Commerce, issued February 2015). Finning itself has been banned since August 2013, and since the ban's implementation, there have been, "no reports of the practice in any part of the country."

2015

Estimating Finite Rate of Population Increase for Sharks
Based on Vital Parameters
PLoS ONE, 10 (11): e0143008
LIU, K.-M. & CHIN, C.-P. & CHEN, C.-H. & CHANG, J.-H.

The vital parameter data for 62 stocks, covering 38 species, collected from the literature, including parameters of age, growth, and reproduction, were log-trasnformed and analyzed using multivariate analyses. Three groups were identified and empirical equations were developed for each to describe the relationships between the predicted finite rates of population increase (λ') and the vital parameters, maximum age (T_{max}), age at maturity (T_m), annual fecundity (f/R_c)), size at birth (L_b), size at maturity (L_m), and asymptotic length (L_∞). Group (1) included species with slow growth rates (0.034 yr^{-1} < k < 0.103 yr^{-1}) and extended longevity (26 yr < T_{max} < 81 yr), e.g., shortfin mako *Isurus oxyrinchus*, dusky shark *Carcharhinus obscurus*, etc.; Group (2) included species with fast growth rates (0.103 yr^{-1} < k < 0.358 yr^{-1}) and short longevity (9 yr < T_{max} < 26 yr), e.g., starspotted smoothhound *Mustelus manazo*, gray smoothhound *M. californicus*, etc.; Group (3) included late maturing species ($L_m/L_\infty \geq 0.75$) with moderate longevity (T_{max} < 29 yr), e.g., pelagic thresher *Alopias pelagicus*, sevengill shark *Notorunchus cepedianus*. The empirical equation for all data pooled was also developed. The λ' values estimated by these empirical equations showed good agreement with those calculated using conventional demographic analysis. The predictability was further validated by an independent data set of three species.

Liu et al. construct a robust analysis of the vital parameters from two stocks of *Carcharhinus leucas*; unfortunately, the authors fail to elaborate on the stocks' locations, and instead provide only their abbreviations (which remain unexplained and unhelpfully linked to general sources like Compagno et al.'s *A Field Guide to Sharks of the World* and the website *FishBase*). SAF could indicate "South Africa", while NGM could likely be "Northern Gulf of Mexico"; these assumptions, however, remain unfortunately unfounded. A summation of vital parameters from both stocks is provided below:

Vital Parameters		*SAF*	*NGM*
Reproductive strategy	**R**	viviparity	viviparity
Size at birth / asymptotic length	L_b/L_∞	0.30	0.24
Size at maturity / asymptotic length	L_m/L_∞	0.84	0.79
Growth coefficient	$k\ (yr^{-1})$	0.0710	0.0760
Maximum age	$T_{max}\ (yr)$	37.07	36.42
Age at maturity	$T_m\ (yr)$	21.00	18.00
Annual fecundity	f/R_c	2.18	2.00
Finite rate of population increase	λ	1.0226 ± 0.0723	1.0351 ± 0.0420
λ estimated from empirical equation	λ'	1.0540	1.0679
$\lambda - \lambda' = |D_i|*$	$|D_i|*$	0.03	0.03

Both SAF and NGM *Carcharhinus leucas* stocks are classified as Group 1: specifically, "slow growing species with high maximum age". Liu et al. found that longevity, age at maturity, and fecundity per year were the most significant parameters for this group, and thus recommend, "protection of adults or TAC management."

2015

The Ecuadorian Artisanal Fishery for Large Pelagics:
Species Composition and Spatio-Temporal Dynamics
PLoS ONE, 10 (8): e0135136
MARTÍNEZ-ORTIZ, J. & AIRES-DA-SILVA, A.M. &
LENNERT-CODY, C.E. & MAUNDER, M.N.

The artisanal fisheries of Ecuador operate within one of the most dynamic and productive marine ecosystems of the world. This study investigates the catch composition of the Ecuadorian artisanal fishery for large pelagic fishes, including aspects of its spatio-temporal dynamics. The analyses of this study are based on the most extensive dataset available to date for this fishery: a total of 106,963 trip-landing inspection records collected at its five principal ports during 2008 – 2012. Ecuadorian artisanal fisheries remove a substantial amount of biomass from the upper trophic-level predatory fish community of the eastern tropical Pacific Ocean. It is estimated that at least 135 thousand metric tons (mt) (about 15.5 million fish) were landed in the five principal ports during the study period. The great novelty of Ecuadorian artisanal fisheries is the "oceanic-artisanal" fleet component, which consists of mother-ship (*nodriza*) boats with their towed fiber-glass skiffs (*fibras*) operating with pelagic longlines. This fleet has fully expanded into oceanic waters as far offshore as 100°W, west of the Galapagos Archipelago. It is estimated that *nodriza* operations produce as much as 80% of the total catches of the artisanal fishery. The remainder is produced by independent *fibras* operating in inshore waters with pelagic longlines and/or surface gillnets. A multivariate regression tree analysis was used to investigate spatio-environmental effects on the *nodriza* fleet (n = 6,821 trips).

The Ecuadorian "oceanic-artisanal" fishery is a unique pelagic innovation, "not known to be widespread elsewhere in the world". Its structure challenges conventional definitions of the small-scale artisanal fishery, and its impact is substantial (primarily when it comes to dolphinfish, *Coryphaena hippurus,* which constitute 64.7% of the oceanic artisanal catch). *Carcharhinus leucas,* however, appears to escape the efficient jaws of this pelagic longline, as it (statistically) constitutes 0.0% of the same catch (equaling to 124 counts of this species, comprising a total of 12.1 metric tons).

It has been a while since our last embrace of the nonempirical; this recent scientific decade is chockablock with Bull Shark statistics, hypotheses, observations and fisheries assessments. But this year—2015— remains in startling contrast to the poetry and adventurism of 1915 (which itself was devoid of the Romantic charm of Müller & Henle's very first publication—the discovery of our beloved *C. leucas*). How have we thus changed? We have become sharper, no doubt; more accurate, more precise, and have decided to waste no time when it comes to the evaluation of this species, its economic role and its fisheries needs. Photographs replace illustrations. Creative prose is no longer appropriate—gone are the days of safely alluding to the Bull Shark as "a monster". Old names—Nicaragua, Cub, *Prionodon, Carcharias*—are all consolidated into a single appropriate identity: *Carcharhinus leucas*—the Bull Shark, FAO code CCE.

But what do we leave behind as we discard the inessential? We, of course, gain so much: the power to understand and protect this species—the power to accurately perceive our physical world. But do we simultaneously lose something meaningful as we shed our chrysalis of unhelpful ideas? It is unquestionably beautiful to emerge into the vibrancy of truth—to bask in its crystalline colors; but is there *meaning* to truth's chrysalis itself?

2015

Spatiotemporal Variability of Fish Assemblage in the
Shatt Al-Arab River, Iraq
Journal of Coastal Life Medicine, 3 (1): 27-34
MOHAMED, A.-R.M. & HUSSEIN, S.A. & LAZEM, L.F.

Objective: To study spatial and temporal variability of fish assemblage in the Shatt Al-Arab River.

Methods: This study was conducted from December 2011 to November 2012. Water temperature, salinity, dissolved oxygen and transparency were measured from three sites in the river. Several fising methods wete adopted to collect fish including gill nets, cast net, electro-fishing and hook and lines. Associations between the distribution of fish species and the environmental variables were quantified by using canonical correspondence analysis.

Results: The results showed that the fish assemblage consisted of 58 species representing 46 genera and 27 families belong to Osteichthyes except one (*Carcharhinus leucas*) relate to Chondrichthyes. Number of species increased in summer and autumn months and sharply decreased in winter. *Tenualosa ilisha* was the most abundant species comprising 27.4% of the catch, followed by *Carassius auratus* (23.7%) and *Liza klunzingeri* (10.6%). The dominance (D_3) value for the main three abundant species was 61.7%. Nine species were caught for the first times from the river include eight marine. The overall values of diversity index ranged from 0.67 in March to 2.57 in October, richness index from 2.64 in January to 3.71 in September and evenness index from 0.22 in March to 0.73 in August.

Carcharhinus leucas is a rare summer visitor to the Shatt Al-Arab (Mohamed et al. report a total relative abundance of 0.004), and has been recorded from both the Tigris and Euphrates: two of the four rivers said to flow from the Biblical Garden of Eden.[13] Of course, the Tigris and Euphrates could be (and perhaps should be) completely assessed without religious context, as true empiricism demands; however, the storied nature of these two rivers—cradles of faith and civilization—begs a fascinating question: what is the Tigris and Euphrates Bull to an Abrahamic theist?

At first, the question may seem silly to ask, as the empirical nature of a Bull Shark does not change between worldviews, and the being itself cannot cater to the designs of a specific practice or culture. However, the significance as *applied* to the Bull Shark may differ widely across beliefs— so much so that the triviality of its existence within both the Tigris and Euphrates watersheds may convey a significance unperceivable by many, but compelling to some. Specifically: if one such Abrahamic theist sincerely believes in the existence of the Garden of Eden—the cradle of man and original sin, allegedly located within the same watershed as the Tigris, Euphrates, Pishon and Gihon rivers—and yet accepts empirical truths (such as Mohamed et al.'s present observation), such a theistic one must entertain the possibility of the Bull Shark's adjacency to Garden itself (and at the dawn of mankind): if this euryhaline species exists in the watershed, what is to have stopped it from entering the Garden's own waters (or from witnessing the fall of man)?

I treat this as a 'thought experiment', and in no way sincerely endorse the concept; yet, it is strangely compelling to entertain—this odd hybridization of true science and religious belief (or fantasy, as alleged). To non-Abrahamics, this must be utter nonsense; but to the Abrahamic theists, could the Bull Shark bear new meaning?

2015

Fishes of the Eastern Johor Strait
Raffles Bulletin of Zoology, Suppl. 31: 303-337
NG, H.H. & TAN, H.H. & LIM, K.K.P. & LUDT, W.B. &
CHAKRABARTY, P.

We record the presence of 435 fish species from the Eastern Johor Strait based on our fieldwork, a review of the existing literature, and an examination of photographs and museum specimens. Four species are recorded for the first time from the waters of Singapore: *Pseudorhombus elevatus* (Paralichthyidae), *Heteromycteris hartzfeldii* (Soleidae), *Nuchequula manusella* (Leiognathidae) and *Johnius carouna* (Sciaenidae).

The Johor Strait (also known as the Straits of Johor and the Tebrau Straits) is the waterway dividing Singapore to the south and Peninsular Malaysia to the north. Approximately 50 km long, the Strait is bisected into a larger eastern and smaller western portion by the Johor-Singapore causeway since its construction in 1923. The Eastern Johor Strait (EJS) run eastwards from the causeway (at 1°27'N 103°46'E) to the easternmost limit of the border between Singapore and Malaysia (at 1°20'N 104°6'E). The EJS is relatively shallow, with the central channel running only about 10—20 m deep in most places. The substrate of the EJS consists largely of unconsolifated sand and mud.

Carcharhinus leucas (Müller & Henle)
Material examined: ZRC 54615

Like Mohamed et al.'s study of the Shatt Al-Arab River, Ng et al.'s *Fishes of the Eastern Johor Strait* explores a relatively uncharted biodiversity. It is rather interesting to consider that many of the coastal areas in Southeast Asia, the Middle East and Oceania are still not well known in terms of their species compositions. Numerous factors come into play in shaping such areas as modern frontiers of biological sampling. Political instability, for example, has doubtlessly limited the opportunities for research surveying in Iraq, and has furthermore introduced unempirical biases into the local understanding of the biological community. One of the most startling examples of this can be found with the 2007 capture of a Bull Shark in the Euphrates River; in the words of a local fisherman on the species' presence, "I believe that America is behind this matter."[14] Disturbingly, a local area teacher corroborated this claim by stating, "there was a 75 percent chance Americans had put the shark in the water."[14]

Apart from the ludicrous amount of academic prohibition that comes with political instability, regional inaccessibility and limited resources (i.e., lack of scientific funding or survey equipment) have likewise hindered the progress of many survey efforts across the globe. Both factors likely play a key role in the prohibition of biological sampling in the islands of Oceania (and possibly some coastal regions in Southeast Asia as well). It is, however, difficult to determine what combinations of factors have led to the sporadic sampling of the Eastern Johor Strait specifically. Unequal research efforts between Singapore and Malaysia could be a possibility, as Ng et al. state that, "records from the Malaysian side of the EJS are largely deficient and the identities of these species recorded in the few inventories available cannot be verified." Of this study's 435 reported fish species, 31 species (7.2%) have, "not been encountered in the last 40 years."

2016

Substitutions in the Glycogenin-1 Gene Are Associated with the Evolution of Endothermy in Sharks and Tunas
Genome Biology and Evolution, in press
CIEZAREK, A.G. & DUNNING, L.T. & JONES, C.S. &
NOBLE, L.R. & HUMBLE, E. & STEFANNI, S. &
SAVOLAINEN, V.

Despite 400-450 million years of independent evolution, a strong phenotypic convergence has occurred between two groups of fish: tunas and lamnid sharks. This convergence is characterised by centralization of red muscle, a distinctive swimming style (stiffened body powered through tail movements) and elevated body temperature (endothermy). Furthermore, both groups demonstrate elevated white muscle metabolic capacities. All these traits are unusual in fish and more likely evolved to support their fast-swimming, pelagic, predatory behaviour. Here we tested the hypothesis that their convergent evolution was driven by selection on a set of metabolic genes. We sequenced white muscle transcriptomes of six tuna, one mackerel and three shark species, and supplemented this data set with previously published RNA-seq data. Using 26 species in total, (including 7,032 tuna genes plus 1,719 shark genes), we constructed phylogenetic trees and carried out maximum-likelihood analyses of gene selection. We inferred several genes relating to metabolism to be under selection. We also found that the same one gene, glycogenin-1, evolved under positive selection independently in tunas and lamnid sharks, providing evidence of convergent selective pressures at gene level possibly underlying shared physiology.

As a Carcharhinid, *Carcharhinus leucas* is not prominently featured in Ciezarek et al.'s study, but instead utilized as a genetic outlier. The origin of utilized *Carcharhinus leucas* samples and their *de novo* trinity assembly statistics are provided as follows:

Origin	SAMN0333334 8
Paired-end reads used for assembly (million)	60.5
Number of assembled contigs	91,122
Contig N50	1,719
Number of coding regions after clustering	21,657

While *Carcharhinus leucas* is of minimal importance to Ciezarek et al. (the species is never once mentioned in the main text), its implicit juxtaposition with the Lamnids makes for an interesting affirmation of the Bull Shark's unique ecological niche and physiological capabilities. The Lamnidae—that is, the White, Salmon, Porbeagle and two Mako Sharks—are all fast, streamlined, torpedo-shaped species capable of rapid-fire attacks from the open sea or coastal shallows. The Bull Shark, by comparison, is more relaxed in its physique; its body, while sleek and swift, isn't as sharply polished as that of the missile-guided Lamnids, and furthermore lacks the critical capacity for countercurrent exchange (via rete mirabilis). This ability to retain heat and raise internal body temperature is essential for rapid muscle contraction and explosive focus: only with such prerequisites can a shark embrace the cold expanse of the open sea and launch itself into fleeting schools of nimble mackerel and dashing tuna for a living. *Carcharhinus leucas* lacks such qualities, and thus could not survive the harsh realities of the open sea for long (likewise, the Lamnids would die in attempting to copy the euryhaline niche of the Bull).

2016

Evidence of Partial Migration in a Large Coastal Predator:
Opportunistic Foraging and Reproduction as Key Drivers?
PLoS ONE, 11 (2): e0147608

ESPINOZA, M. & HEUPEL, M.R. & TOBIN, A.J. & SIMPFENDORFER, C.A.

Understanding animal movement decisions that involve migration is critical for evaluating population connectivity, and thus persistence. Recent work on sharks has shown that often only a portion of the adult population will undertake migrations, while the rest may be resident in an area for long periods. Defining the extent to which adult shark use specific habitats and their migratory behaviour is essential for assessing their risk of exposure to threats such as fishing and habitat degradation. The present study used acoustic telemetry to examine residency patterns and migratory behaviour of adult bull sharks (*Carcharhinus leucas*) along the East coast of Australia. Fifty-six VR2W acoustic receivers were used to monitor the movements of 33 bull sharks in the central Great Barrier Reef (GBR). Both males and females were detected year-round, but their abundance and residency peaked between September and December across years (2012-2014). High individual variability in reef use patterns was apparent, with some individuals leaving the array for long periods, whereas others (36%) exhibited medium (0.20-0.40) or high residency (> 0.50). A large portion of the population (51%) undertook migrations of up to 1,400 km to other coral reefs and/or inshore coastal habitats in Queensland and New South Wales. Most of these individuals (76%) were mature females, and the timing of migrations coincided with the austral summer (Dec-Feb).

The Great Barrier Reef (specifically the central area encompassing the Townsville Reefs, approximately 70 km offshore from Townsville, Queensland) appears to play a much greater role in the foraging and migration behavior of Australian Bull Sharks than previously imagined; in fact, Espinoza et al. go as far as to claim that, "bull sharks may play an important role as top predator in this ecosystem". The Townsville Reef area is characterized by, "the presence of semi-isolated midshelf reefs (5-15 km apart) separated by relatively deep sandy channels (40-70 m)." Spanish Mackerel (*Scomberomorus commerson*) are known to spawn here from September to November and form, "large predictable aggregations in some of the midshelf reefs." The Bull Shark evidently spends a disproportionate amount of time at these same reefs in conjunction with the Mackerel spawning season, and as a potential foraging strategy, concurrently utilizes, "movement corridors along inner midshelf reefs within this region."

A large number of female *Carcharhinus leucas* leave the central Great Barrier Reef area in the late spring and early summer. Within this particular study, a small portion of tagged migratory females were detected by nearshore arrays for relatively short periods of time (less than 5 days on average) and in synchrony with the species' reported parturition; these observations have particularly compelled Espinoza et al. to suggest that, "(i) inshore movements were potentially to give birth; (ii) female reproductive migrations may be a major driver influencing seasonal patterns of reef use and population structure along the East coast of Australia; and (iii) females may return to their birth place and/or birth region for parturition." On this third point concerning reproductive philopatry, the authors specifically allude to a study by Tillett et al. (2012), which suggests (on the basis of mtDNA) that, "females may use specific nurseries in multiple breeding events."[15]

2016

New Record of Carcharhinus leucas (Valenciennes, 1839)
in an Equatorial River System
Marine Biodiversity Records, 9: 87
FEITOSA, L.M. & MARTINS, A.P.B. & NUNES, J.L.S.

Bull sharks are a cosmopolitan shark species frequently found in shallow shelf ocean waters and, occasionally, in several tropical river systems around the world. Due to the bull shark's capability to enter riverine systems, the documentation of its occurrence is essential for future fisheries inspections and studies. In this way, this study aims to report the presence of a medium sized specimen of *C. leucas* in an equatorial river system. The specimen was caught by fishermen at Mearim River, located in Northern Brazil and well known for the occurrence of tidal bores during the highest spring tides of the dry season. The event coincided with the occurrence of one of the strongest spring tides of 2015. The captured female specimen measured approximately 1300 mm and weighted 35 kg. The occurrence of this species was not known in this river basin until now. We recommend and support future ichthyologic studies in the Mearim River basin in order to provide data for the delimitation of the territory used by *C. leucas* in Maranhão State, specially looking into its age, growth, diet, spatial, and temporal movement patterns in this area.

Keywords
Euryhaline shark Elasmobranch   North Brazil

The Mearim River is most famous for its *pororoca*—the tidal bores that occur from August to December and raise salinities to a much as 18 ppt. The extent of the *pororoca* can be felt as far upriver as Arari city, while tidal influence itself projects as much as 256 km into the basin. Slow currents, muddy sediments, turbid waters and rich nutrients characterize the lower portions of the Mearim River (which spans a total length of 930 km and empties into the, "meridional edge of *Ilha dos Caranguejos*"). It is in this setting that an immature female Bull Shark was captured; according to its captors, the Bull Shark "tried to attack a dog on the margin and got stranded due to the low depth."

Feitosa et al. affirm that only 5% of all elasmobranchs can, "tolerate some sort of salinity range during their lifetime." In the case of the truly euryhaline *Carcharhinus leucas*, the mechanism behind this tolerance is rather complicated, and can be a challenge to succinctly comprehend. Feitosa et al. nevertheless take aim to neatly explain the elaborate brilliance behind the Bull Shark's osmoregulatory prowess: as presented by the authors, the Bull Shark's osmotic acclimation to salinity gradients is, "believed to be related to the rectal gland activity plasticity, urea and trimethylamine oxide (TMAO) reabsorption by the kidney, and ion uptake by the gills, which is enhanced when in freshwater." The combined effect of these processes results in a Bull Shark hyperosmotic to its freshwater surroundings, but consequentially prone to the loss of, "much more water and ions due to large amounts of urine produced." By similar (but less efficient) means, other elasmobranchs have wandered into the estuarine areas of Maranhão State: these include *Isogomphodon oxyrhynchus* (Daggernose Shark), *Carcharhinus porosus* (Smalltail Shark), *Sphyrna tiburo* (Bonnethead Shark), *Rhizoprionodon porosus* (Caribbean Sharpnose Shark) and *Pristis pristis* (Largetooth Sawfish).

2016

Cyanobacterial Neurotoxin BMAA and Mercury in Sharks
Toxins, 8 (8): 238
**HAMMERSCHLAG, N. & DAVIS, D.A. & MONDO, K. &
SEELY, M.S. & MURCH, S.J. & GLOVER, W.B. & DIVOLL,
T. & EVERS, D.C. & MASH, D.C.**

Sharks have greater risk for bioaccumulation of marine toxins and mercury (Hg), because they are long-lived predators. Shark fins and cartilage also contain β-N-methylamino-L-alanine (BMAA), a ubiquitous cyanobacterial toxin linked to neurodegenerative diseases. Today, a significant number of shark species have found their way onto the International Union for Conservation of Nature (IUCN) Red List of Threatened Species. Many species of large sharks are threatened with extinction due in part to the growing high demand for shark fin soup and, to a lesser extent, for shark meat and cartilage products. Recent studies suggest that the consumption of shark parts may be a route to human exposure of marine toxins. Here, we investigated BMAA and Hg concentrations in fins and muscles sampled in ten species of sharks from the South Atlantic and Pacific Oceans. BMAA was detected in all shark species with only seven of the 55 samples analyzed testing below the limit of detection of the assay. Hg concentrations measured in fins and muscle samples from the 10 species ranged from 0.05 to 13.23 ng/mg. These analytical test results suggest restricting human consumption of shark meat and fins due to the high frequency and co-occurrence of two synergistic environmental neurotoxic compounds.

Three Bull Sharks were sampled from Florida Bay and tested for BMAA, total mercury and methyl mercury concentrations. The collective results are as follows:

Range (BMAA)	43-264 ($n = 3$) ng/mg
Detected Mean ± SE	180 ± 69 ($n = 3$) ng/mg
BMAA/Length	103 ng/100 ($n = 3$) cm
Range (Mercury)	3.24-13.23 ng/mg
Total Mercury (THg)	7.26 ± 3.04 ($n = 3$) ng/mg
Methyl Mercury (MeHg)	2.32 ($n = 1$) ng/mg
BMAA: THg	27:1

While (wet weight) BMAA concentrations in *Carcharhinus leucas* were below average (180 ± 69 ng/mg, in comparison with a survey mean of 366 ± 72 ng/mg), total mercury concentrations were exceptional; *Carcharhinus leucas* possessed the highest mercury levels, with a maximum THg concentration of 13.23 ng/mg and a mean THg concentration of 7.26 ± 3.04 ng/mg (survey mean = 2.3 ± 0.4 ng/mg). This result is not surprising, given that the Bull Shark is an apex predator and that mercury bioaccumulates at the highest trophic levels; however, this species' mercury concentrations are, "higher than those reported safe for human consumption, which range from 0.3 to 1.0 μg / g wet weight." Indeed, the Florida Department of Health (FDOH) advises that, "people should not eat sharks greater than ~ 109 cm". Hammerschlag et al. express specific concern for neurotoxic BMAA exposure in Florida waters, and refer to the anticipated increase of Alzheimer's disease (a potential consequence of the combined effects of mercury and BMAA) as a possible correlation (from 0.5 to 0.7 million senior residents by 2020).

2016

A Checklist of Helminth Parasites of Elasmobranchii in Mexico

Mexico

Zookey, 563: 73-128

IVAN MERLO-SERNA, A. & GARCIA-PRIETO, L.

A comprehensive and updated summary of the literature and unpublished records contained in scientific collections on the helminth parasites of the elasmobranchs from Mexico is herein presented for the first time. At present, the helminth fauna associated with Elasmobranchii recorded in Mexico is composed of 132 (110 named species and 22 not assigned to species), which belong to 70 genera included in 27 families (plus 4 *incertae sedis* families of cestodes). These data represent 7.2% of the worldwide species richness. Platyhelminthes is the most widely represented, with 128 taxa: 94 of cestodes, 22 of monogeneans and 12 of trematodes; Nematoda and Annelida: Hirudinea are represented by only 2 taxa each. These records come from 54 localities, pertaining to 15 states; Baja California Sur (17 sampled localities) and Baja California (10), are the states with the highest species richness: 72 and 54 species, respectively. Up to now, 48 elasmobranch species have been recorded as hosts of helminths in Mexico; so, approximately 82% of sharks and 67% of rays distributed in Mexican waters lack helminthological studies. The present list provides the host, distribution (with geographical coordinates), site of infection, accession number in scientific collections, and references for the parasites. A host-parasite list is also provided.

Ivan Merlo-Serna & Garcia-Prieto record nine species of parasitic helminthes from the spiral valve of *Carcharhinus leucas* (details are provided below). The associated specimens are currently housed in four collections: the Colección Nacional del Helmintos (CNHE); the Lawrence R. Penner Collection (LRP); the United States National Parasite Collection (USNPC); and the Connecticut State Museum of Natural History (CSMNH).

Species	Locality	Collection
Cathetocephalus resendezi	Bahía de Los Ángeles, BCN	CNHE (5300)
Cathetocephalus thatcheri	Playa Chachalacas, VER	CNHE (6860)
Callitetrarhynchus gracilis	Playa Chachalacas, VER	CNHE (6867)
Eutetrarhynchidae gen. sp.	Playa de Chachalacas, VER	CNHE (6169)
Phoreiobothrium sp.	Bahía de Los Ángeles, BCN	CNHE (6866)
Phoreiobothrium sp.	Playa Chachalacas, VER	CNHE (6866)
Platybothrium angelbahiense	Bahía de Los Ángeles, BCN	CNHE (4727-9)
Platybothrium angelbahiense	Bahía de Los Ángeles, BCN	LRP (3213-3215)
Platybothrium angelbahiense	Bahía de Los Ángeles, BCN	USNPC (92236)
Platybothrium sp.	Bahía de Los Ángeles, BCN	CSMNH
Otobothrium sp.	Playa Chachalacas, VER	CNHE (6863-3)
Paraorygmatobothrium sp.	Playa Chachalacas, VER	CNHE (6864-5)

2016

The Path towards Endangered Species: Prehistoric Fisheries in Southeastern Brazil
Plos One, 11 (6): e0154476
LOPES, M.S. & BERTUCCI, T.C.P. & RAPAGNA, L. & TUBINO, R.D. & MONTEIRO-NETO, C. & TOMAS, A.R.G. & TENORIO, M.C. & LIMA, T. & SOUZA, R. & CARRILLO-BRICENO, J.D. & HAIMOVICI, M. & MACARIO, K. & CARVALHO, C. & SOCORRO, O.A.

Brazilian shellmounds are archaeological sites with a high concentration of marine faunal remains. There are more than 2000 sites along the coast of Brazil that range in age from 8,720 to 985 cal BP. Here, we studied the ichthyoarchaeological remains (i.e., cranial/postcranial bones, otoliths, and teeth, among others) at 13 shellmounds on the southern coast of the state of Rio de Janeiro, which are located in coastal landscapes, including a sandy plain with coastal lagoons, rocky islands, islets and rocky bays. We identified patterns of similarity between shellmounds based on fish diversity, the ages of the assemblages, littoral geomorphology and prehistoric fisheries. Our new radiocarbon dating, based on otolith samples, was used for fishery characterization over time. A taxonomical study of the ichthyoarchaeological remains includes a diversity of 97 marine species, representing 37% of all modern species (i.e., 265 spp.) that have been documented along the coast of Rio de Janeiro state. This high fish diversity recovered from the shellmounds is clear evidence of well-developed prehistoric fishery activity that targeted sharks, rays and finfishes in a productive area influenced by coastal marine upwelling.

The Ilha do Cabo Frio shellmound is, "located on a small sandy beach characterized by an active dune that faces towards the landscape, associated with the outcrop layers that overlap the Cabo Frio beach rock." Ranging in age from 2710 to 3290 cal BP (calendar years Before Present), the Ilha do Cabo Frio shellmound is the only site in which *Carcharhinus leucas* teeth were found (within the scope of this study). These ancient Bull Shark teeth possessed drilled perforations at the root. Such bores could be indicative of ceremonial use (i.e., shark tooth necklaces used for burial rituals or other religious purposes), or, alternatively, for, "tool manufacturing, e.g., affixed to pieces of wood with vegetal fibers, similar to Polynesian artifacts, or as points of arrows."

The presence of drilled teeth from not only *Carcharhinus leucas*, but also: *C. altimus* (Bignose Shark); *C. plumbeus* (Sandbar Shark); *Galeocerdo cuvier* (Tiger Shark); *Sphynra mokarran* (Great Hammerhead) and *Carcharodon carcharias* (White Shark) is testament to a well-developed fishery capable of pelagic exploitation. While early shellmounds (dating from 5,595 cal BP) were characterized by catches of coastal species associated with sandy bottoms and nearshore lagoons (such as croaker, catfish, and mullet), later shellmounds (dating from 4,414 cal BP) incorporated a much broader pelagic diversity of species associated with protected rocky bays and rocky islets. By 3,290 cal BP, advances in multi-gear and artisanal fishing techniques allowed indigenous fishermen to exploit pelagic resources associated with oceanic islands (such as Cabo Frio and Ilha Grande). These latter two areas could have been part of a nursery complex including Arraial do Cabo and Angra dos Reis, as their collections of small teeth and vertebrae from Sphyrnid (hammerhead), Carcharhinid (requiem) and Lamnid (mackerel) sharks provides, "irrefutable support for the presence of pregnant females and juveniles".

2016

Zoogeography of Elasmobranchs in the Colombian Pacific Ocean and Caribbean Sea
Neotropical Ichthyology, 14 (2): e140134
NAVIA, A.F. & MEJÍA-FALLA, P.A. & HLEAP, J.S.

In order to investigate zoogeographical patterns of the marine elasmobranch species of Colombia, species richness of the Pacific and Caribbean and their subareas (Coastal Pacific, Oceanic Pacific, Coastal Caribbean, Oceanic Caribbean) was analyzed. The areas shared 10 families, 10 genera and 16 species of sharks, and eight families, three genera and four species of batoids. Carcharhinidae had the highest contribution to shark richness, whereas Rajidae and Urotrygonidae had the greatest contribution to batoid richness in the Caribbean and Pacific, respectively. Most elasmobranchs were associated with benthic and coastal habitats. The similarity analysis allowed the identification of five groups of families, which characterize the elasmobranch richness in both areas. Beta diversity indicated that most species turnover occurred between the Coastal Pacific and the two Caribbean subareas. The difference in species richness and composition between areas may be due to vicariant events such as the emergence of the Isthmus of Panama. It is unlikely that the Colombian elasmobranch diversity originated from a single colonization event. Local diversification/speciation, dispersal from the non-tropical regions of the Americas, a Pacific dispersion and an Atlantic dispersion are origin possibilities without any of them excluding the others.

Three prominent hypotheses are offered to explain the biogeographical origin of sharks and rays in Colombia (and other parts of the Neotropical realm). Firstly, the Tethyan hypothesis proposes that, "the range of coastal species was sundered by closure of the Tethys Sea, linking the Indian and Atlantic Ocean coasts and isolating American and western Atlantic species." Under this hypothesis, the subsequent emergence of the Panama Isthmus further isolated American Atlantic elasmobranchs from those of the American Pacific. This possibility quite nicely fits with the second hypothesis of local speciation, wherein, "amphi-American genera with species endemic to the neotropics (*e.g. Sphyrna tudes* vs. *Sphyrna media, Squatina dumeril* vs. *Squatina californica*) follow a germinate species pattern likely split by the Isthmus of Panama." The third hypothesis (specific to the Rajidae, or Skates) relates to a broader potentiality of Austral origin via Gondwanan speciation (before the separation of Australia and South America); prior studies have proposed that, "some of the Rajidae subfamilies existed in Gondwana before its fragmentation, giving to the austral region both palaeo-Austral and Tethyan remnants, and therefore a greater diversity." However, Navia et al. question this last position by alternatively proposing that, "the scant presence of the species in the Atlantic and Indian Oceans suggest that a Tethyan link is unlikely and therefore, Rajidae representation in Colombian waters is more likely due to dispersal from southern South America."

All three hypotheses may lend to the extensive biogeographical presence of *Carcharhinus leucas* in Colombian waters (the species can be found in each of the four subregions as defined by Navia et al.) Indeed, the Carcharhinidae as a family contributed the most to total Colombian shark biodiversity (15.38%), and furthermore dominated the relative contributions (per family) for both the Caribbean (17.33%) and Pacific (21%).

2016

Isolation and Characterization of Eight Microsatellite Loci from Galeocerdo cuvier (Tiger Shark) and Cross-Amplification in Carcharhinus leucas, Carcharhinus brevipinna, Carcharhinus plumbeus and Sphyrna lweini

PeerJ , 4: e2041

PIROG, A. & JAQUEMET, S. & BLAISON, A. & SORIA, M. & MAGALON, H.

The tiger shark *Galeocerdo cuvier* (Carcharhinidae) is a large elasmobranch suspected to have, as other apex predators, a keystone function in marine ecosystems and is currently considered Near Threatened (Red list IUCN). Knowledge on its ecology, which is crucial to design proper conservation and management plans, is very scarce. Here we describe the isolation of eight polymorphic microsatellite loci using 454 GS-FLX Titanium pyrosequencing of enriched DNA libraries. Their characteristics were tested on a population of tiger shark ($n = 101$) from Reunion Island (South-Western Indian Ocean). All loci were polymorphic with a number of alleles ranging from two to eight. No null alleles were detected and no linkage disequilibrium was detected after Bonferroni correction. Observed and expected heterozygosities ranged from 0.03 to 0.76 and from 0.03 to 0.77, respectively. No locus deviated from Hardy-Weinberg equilibrium and the global F_{IS} of the population was of 0.04^{NS}. Some of the eight loci developed here successfully cross-amplified in the bull shark *Carcharhinus leucas* (one locus), the spinner shark *Carcharhinus brevipinna* (four loci), the sandbar shark *Carcharhinus plumbeus* (five loci) and the scalloped hammerhead shark *Sphyrna lewini* (two loci).

Microsatellites are tracts of 'repetitive DNA' characterized by their unique (and repeating) sequence motifs. These tracts are excellent tools for population genetics and DNA profiling; as such, knowledge of their locations within an organism is essential for critical population analyses. Prior to this study, only twelve microsatellite loci were available for *Galeocerdo cuvier*; of these, three loci were initially characterized from *Carcharhinus leucas* (in a previous study by the same authors[17]) but were found to be polymorphic for the Tiger Shark as well. This current endeavor by Pirog et al. yields eight new microsatellite loci for *Galeocerdo cuvier* (bringing the total number of available loci up to twenty); of these, only one locus—Gc01—can be successfully cross-amplified in *Carcharhinus leucas*. The polymorphic Gc01 contains three alleles and ranges in size from 136 to 142 base pairs. Within this study, Pirog et al. observed the amplification of Gc01 forty-one times (from a sample of forty-one Bull Sharks). The characterization of Gc01 is provided in the following table (note that the annealing temperature $Ta = 55°C$ and that no locus deviated from Hardy-Weinberg equilibrium):

Locus name	Gc01
Repeat motif	$(GA)_6$
Forward primer (5'-3')	AGGTGTGGTGGCTCTCCTC
Reverse primer (5'-3')	GGACGCAAAATCCAACAGAG
Allele size (bp)	143-147
Number of alleles per locus (N_a)	2
Observed heterozygosity (H_O)	0.03
Expected heterozygosity (H_E)	0.03
Inbreeding coefficient (F_{IS})	- 0.01

2016

Patterns of Occurrence of Sharks in Sydney Harbour, a Large Urbanised Estuary

PLoS ONE, 11 (1): e0146911

SMOOTHEY, A.F. & GRAY, C.A. & KENNELLY, S.J. & MASENS, O.J. & PEDDEMORS, V.M. & ROBINSON, W.A.

Information about spatial and temporal variability in the distribution and abundance of shark-populations are required for their conservation, management and to update measures designed to mitigate human-shark interactions. However, because some species of sharks are mobile, migratory and occur in relatively small numbers, estimating their patterns of distribution and abundance can be very difficult. In this study, we used a hierarchical sampling design to examine differences in the composition of species, size- and sex-structures of sharks sampled with bottom-set longlines in three different areas with increasing distance from the entrance of Sydney Harbour, a large urbanized estuary. During two years of sampling, we obtained data for four species of sharks (Port Jackson, *Heterodontus portujacksoni*; wobbegong, *Orectolobus maculatus*; dusky whaler, *Carcharhinus obscurus* and bull shark, *Carcharhinus leucas*). Only a few *O. maculatus* and *C. obscurus* were caught, all in the area closest to the entrance of the Harbour. *O. maculatus* were caught in all seasons, except summer, while *C. obscurus* was only caught in summer. *Heterodontus portusjacksoni* were the most abundant species, caught in the entrance location mostly between July to November, when water temperature was below 21.5°C. This pattern was consistent across both years.

A total of twelve *Carcharhinus leucas* were acoustically tagged within the two-year span of Smoothey et al.'s study. The Bull Shark's low relative abundance in Sydney Harbour is unsurprising as, "one may expect from a relatively rare, mobile species at the southern extent of its distribution." Smoothey et al. determine a sensible temperature range for this species in Sydney Harbour, stating that the Bull Shark was very unlikely to be caught (P | capture | < 0.02) below 19°C and more likely to be caught than not (P | capture | > 0.50) at about 23.2°C. These temporal patterns may correspond to the seasonal shift of the East Australian Current (EAC); as speculated by the authors, *C. leucas* may, "use the strong EAC to migrate southwards during summer and return northward during winter, when the current is weakest and inshore waters are cooler." Both migration patterns (which never resist the current, and theoretically match the Bull Shark's optimal temporal preferences) may be in attempt to, "minimize energetic demands of key metabolic and physiological processes". Alternatively, Smoothey et al. speculate that *C. leucas* may be *indirectly* responding to the area's change in temperature—that the Bull Shark may be simply mirroring the temporal responses of essential prey items such as yellowtail kingfish (*Seriola lalandi*), Australian bonito (*Sarda australis*), frigate mackerel (*Auxis thazard*) and mackerel tuna (*Euthynnus affinis*).

As most of the twelve captured Bull Sharks were adults (with one bearing fresh mating scars), Smoothey et al. ponder the possibility of Sydney Harbour as a potential summer mating ground for *Carcharhinus leucas*. Theoretically, mating would occur in January: the time of peak abundance for sexually mature individuals. This possibility neatly coincides with the facts that neonates can be found in the northern rivers of New South Wales in November and that the species' gestation period lasts from 10-11 months.

2016

**Evidence of Positive Selection Associated with
Placental Loss in Tiger Sharks**
BMC Evolutionary Biology, 16 (1): 126
SWIFT, D.G. & DUNNING, L. & IGEA, J. & BROOKS, E.J. &
JONES, C.S. & NOBLE, L.R. & CIEZAREK, A. & HUMBLE,
E. & SAVOLAINEN, V.

All vertebrates initially feed their offspring using yolk reserves. In some live-bearing species these yolk reserves may be supplemented with extra nutrition via a placenta. Sharks belonging to the Carcharhinidae family are all live-bearing, and with the exception of the tiger shark (*Galeocerdo cuvier*), develop placental connections after exhausting yolk reserves. Phylogenetic relationships suggest that lack of placenta in tiger sharks is due to secondary loss. This represents a dramatic shift in reproductive strategy, and is likely to have left a molecular footprint of positive selection within the genome.

We sequenced the transcriptome of the tiger shark and eight other live-bearing shark species. From this data we constructed a time-calibrated phylogenetic tree estimating the tiger shark lineage diverged from the placental carcharhinids approximately 94 million years ago. Along the tiger shark lineage, we identified five genes exhibiting a signature of positive selection. Four of these genes have functions likely associated with brain development (*YWHAE* and *ARL6IP5*) and sexual reproduction (*VAMP4* and *TCTEX1D2*).

Our results indicate the loss of placenta in tiger sharks may be associated with subsequent adaptive changes in brain development and sperm production.

Although Swift et al. primarily focus on *Galeocerdo cuvier* within their study, the incorporation of *Carcharhinus leucas* yields compelling insight in regards to the Bull Shark's evolution and use of placental viviparity as an adaptation. After sequencing the transcriptomes of *Carcharhinus leucas*, *Galeocerdo cuvier* and seven other shark species, Swift et al. identified 3,215 putative orthologous sequences, which were subsequently filtered (via the generation of confidence alignments, and the selection against low-confidence codon alignments) into 1,197 orthologues suitable for further analysis. These 1,197 orthologue alignments were then, "concatenated (1,101,288 bp) and used to construct a phylogenetic tree using RAxML (version 8.0.0)."

Swift et al. determined that 95 of the 1,197 orthologues were, "found to have specific codon sites showing signatures of positive selection". From the remaining 1,102 orthologues (not evolving under positive selection), the authors reconstructed a second phylogenetic tree (typologically identical to the first). To this tree, Swift et al. applied fossil data to obtain a time-calibrated phylogeny of the nine shark species. As a result of this entire process, Swift et al. elucidated that the lineage containing *Carcharhinus leucas* and *Carcharhinus acronotus* (which potentially encompasses the entire 'Ridgeless *Carcharhinus*' subgroup) evolved between 11 and 22 million years ago (mean = 17 mya).

C. leucas exhibits placental viviparity—a trait thought to have, "one evolutionary origin in sharks." Only the Carcharhinidae, Sphyrnidae, Hemigaleidae, Leptochariidae, and Triakidae are placentally viviparous. This adaptation is suggested to, "increase the embryonic development rate of energetically expensive tissues such as the brain." Indeed, placental species like *Carcharhinus leucas* are thought to have higher brain-to-body mass ratios than aplacental sharks (possibly as a result of this adaptation).

2016

Vertebrados Marinos del Neógeno del Suroeste de la Península Ibérica
Thesis, University of Huelva, Spain
TOSCANO-GRANDE, A.

This PHD thesis studies the marine vertebrates present in the Miocene and Pliocene formations of Southwestern Guadalquivir Basin in the province of Huelva, its evolutionary aspects and paleoecological implications. The result obtained allow to define with higher resolution the development of coastal and marine ecosystems in the Southwest of the Iberian Peninsula during the late Neogene. During this time, the two main pathways between the Atlantic Ocean and the Mediterranean Sea were Norbético Strait and the Strait Rif. The studied area for their paleogeographic situation between Atlantic and Mediterranean is in a prime location for the study of the fossil record of these groups.

The Neogene studied materials belong to the following formations: Fm. Niebla (Tortoniense sup.) Fm. Arcillas de Gibraleón (Tortoniense sup.-Messinian), and Fm. Arenas de Huelva (Pliocene inf.). Representative sites were selected and was made systematic sampling (350 kg.), and surface searches. Also, a facies mapping and geological sections ware made. The samples were sieved in the laboratory in order to separate the remains of small vertebrates, and microfossils to detail the paleoecological conditions and their dating. They also cleaned and consolidated large vertebrate fossils. The remains were cataloged, laveled and photographed.

Toscano-Grande collected 112 *Carcharhinus leucas* teeth from sixteen deposits ranging in age from the Upper Miocene (6-8 Ma) to the present. Most teeth originated from the Arenas de Huelva (Lower Pliocene; nearshore environment, 30-100m deep) and Arcillas de Gibraleón (Upper Tortonian-Messinian; offshore environment, 150-400m deep) formations. The specific distributional record for the Ancient Bull is as follows:

Formation	Deposit	Frequency
HUELVA	*Marismas de Mendaña*	Very Frequent (>21)
HUELVA	*Marismas del Rincón-Vias*	Scarce (4-10)
HUELVA	*Marismas del Rincón-Fosfoyesos*	Very Scarce (1-3)
HUELVA	*Marismas del Polvorín Norte*	Common (11-15)
HUELVA	*Marismas del Polvorín Sur*	Frequent (16-20)
HUELVA	*Avenida de Cádiz*	Common (11-15)
HUELVA	*Cabezo de la Plaza de Toros*	Very Frequent (>21)
HUELVA	*Cabezo de la Fábrica de Harina*	Scarce (4-10)
GIBRALEÓN	*Autovía de Gibraleón*	Frequent (16-20)
TRIGUEROS	*Cruce de Trigueros*	Very Scarce (1-3)
BONARES	*Bonares- Ambulatorio*	Frequent (16-20)
BONARES	*Bonares- Km 8*	Very Scarce (1-3)
PALOS DE LA FRONTERA	*Palos- Camino del Tubo Amarillo*	Very Scarce (1-3)
CORRALES	*Santa Clara 1*	Very Frequent (>21)
CORRALES	*Santa Clara 2*	Common (11-15)
CARTAYA	*Nuevo Portil*	Very Scarce (1-3)

Thus concludes the scientific compendium—the complete collection of all publically available empirical works on the Bull Shark, Carcharhinus leucas, from 1841 to 2016.

However, our journey is not yet over...

We have but one final stage. Throughout the entirety of this work, we have infused some elements of the supranatural—the aesthetic, the mythological, and the Romantic—into the scientific presentation of *Carcharhinus leucas*. Before now, we have kept such supranaturality at a comfortable distance from our empirical foundations—the skeleton of *understanding*...but now we must come to our final point: a place sitting between the edge of reason and the unmarkable beyond. This is not an easy place to find, especially if one has the tendency to overthink—to become locked within the rules of language and logic...

But such a place shall be the soul of our *understanding*: this is the *Meditatio Penultima*—the Penultimate Meditation. It is significantly 'Penultimate', because the final meditation— *ultimate understanding*—lies only with you. Only you can discern the true sense of meaning from this account of *Carcharhinus leucas;* you have the exclusive power to accept or reject what arguments, ideas or meditations I may present in relation to this beautiful shark. The *Meditatio Penultima* will be what you make of it; my only hope is that it may perhaps be a primer for profound contemplation.

Let us now gaze inward, beyond the threshold...we are at the doorstep of a Museum—the Museum that we have so diligently constructed from 175 years of thought and 194 works of empiricism. The touch of thirty-one cultures graces the walls and halls of this penultimate place, and compliments its structure with motifs of professional science and personal taste. But now, deep within this marvelous Musuem, there exists something far more powerful than anything that we, or any other, can create with thought alone...the Portrait within—the Portrait of *Carcharhinus leucas*. We must find this...and after a few deep and comfortable breaths...

∞

We must enter.

MEDITATIO PENULTIMA

Carcharhinus leucas is the title on the plaque. The plaque is brassy, polished, but old; it adorns the base of a gigantic frame—twelve feet tall and twenty-four feet wide (an unfairly constructed figure, born from my American eyes)—precisely centered, yet comfortably so at a complimentary size. Many adornments are hewn, cut, etched, chiseled, painted, pasted, layered and plastered upon the Frame. Both wings of the Frame—left, and right—exhibit human bones. Bronze femurs gently curve upwards upon other limbs, but the Frame is absent of skulls, ribs, hands and feet. Teeth marks are etched into each bone, but these are not frightening; they are heavy, and more quieting than disquieting—silence is of utmost importance when examining the Frame.

Blood and testosterone are tiny features—round specks, like red sands in an hourglass, only softer in shape and more precious—and these tumble in and out of the many nooks and crannies of the Frame, between bones, serrations and motifs of sand. A tooth adorns each of the Frame's four corners: the top left and top right, 5th position, and bottom left and bottom right, 4th position. The smaller, pyramidal shapes curve upward and perfectly inward into the Portrait itself, whilst the larger, commanding, dangerous but bewitching uppers protect the integrity of the portrait within, and defend downward against all molesters of the aesthetic beauties of the world. Sand and mud smother the Frame's bottom—suspended sediment mixing with the red spheres of blood and testosterone (I admit, this Frame is not the most exciting piece to behold) yet provide balance and security against the more chaotic elements. Symbols of gods can be found if you look close enough: an X for Xipactli, S for Sura, Y for Eden, and an inward spiral for the Dreamtime (these, however, are diminutive and scattered, and only deserving of the keenest eye). When you shift your position, the overwhelming yet comfortable amount of bronze shines with a brilliant scattering of bright lazuli—concealed within the Frame's more secretive designs. Indeed, when you look even more closely, you can see an olive sheen accentuating the magnificence with its subtle background—like stars, the tiny blue speckles bewilder and delight. The Frame's ugliness becomes more complex.

Mud, blood and bones are a bit less important. You learned that the Frame itself was cooled by pure freshwater, but that doesn't really concern you much. Salt has gently calcified the Frame's stronger elements, but again, that matters little. Tropicality isn't found in the Frame, though it should be (then again, you could always move the entire Portrait to whatever venue you desire—Mexico, Guatemala, Nicaragua, Cuba, Germany, Paris, Italy, Guinea, Azores, South Africa, Pakistan, Iraq, India, Indonesia, Singapore, Malaysia, Vietnam, Cambodia, Australia, Peru, the United States—it doesn't matter). It really doesn't matter. We are proud of the Frame Makers. We admire their work. And they hold the Portrait into shape, and accentuate it with precisely chosen motifs: Near Threatened; fisheries rarity; FAO CCE; Bull Shark; *Carcharhinus leucas*; ridgeless; bold; dangerous; euryhaline; 150m depth limit; neritic; intertidal; coral reef. The Bull Shark's tail pulls precise vortices out of chaos and into existence. Barium Ferrite pushes precise frequencies into the Bull Shark's senses and overloads them for our protection. We admire the Frame Makers—we *need* the Frame Makers, to protect the Portrait within. The shell, it is. The portal to our world, as it acts. Its purpose in our gallery is to bridge our minds. But...

Look at the Portrait

Look

Clear blue. No animals. The sky is too far above; this is below. Beneath the meniscus of the sea, where sunlight penetrates but air does not. The medium is blue. Gradients are measured in blue. Dark blue is far to the West—too far to swim. A whale passing by would disappear as a speck into that kind of dark blue. Its darkness lies in between the clear blue and the blueing white of the sands. No horizon can be seen so far away; dark blue is eternal blue. But some blue is closer—surface blue and clear blue mix in colorful harmony upwards in warmer, more intimate waters. Surrounding clear blue and surface blue mix and mingle and lap and sing that irreplaceable bass note of lapping undersea waters and waves. The medium bass—medium encompassing everything in comfortable, intimate, warm immediacy. Sunlight firing from the air curls into this kind of blue, gently warming the surface and clearness, in cooled ferocity. It dances in the water like a nymph—unexpectedly pondered, but unquestionably playful. Sunlight gives clear blue its color and light. Light dances in a thousand waves and ripples. The disappearing act is better than any magic. Light is playful, curling, dancing, considerate.

The sands at the bottom are white and bright with smooth, tiny grains, gently caressed into scoops and slopes. The White Sahara in all of its diminutiveness, all of its secrecy, and all of its exclusivity extends at the bottom; white sands, receptive of bottom blue, receptive of the bottom light. A pink starfish daring to enter the Portrait may still retain its gentle hue. It is not so deep, but deep enough for bottom blue and dark blue in the distance. Shallow blue, however, mixes with clear blue and surface blue, all passing light. Bending, curling beams are gently encouraged to arc from playful blue to playful blue. Emptiness is impossible, and not because of the animal suspended between all of these blues and lights and sands and waves and tides. It's because the others exist, bending, arcing, relaxing, playing, shining, sleeping, surfing, sailing, and swimming.

But an animal is here too. She is in the center of this scene, but not because she is admired by the blues and the lights and the sands and the tides. She is one of them—the same, rearranged into a magnificent complexity more tangible than the rest. She is massive and heavy, but acts as light as the clear blue, the surface blue, and the shallow blue: this is one of her many most magnificent tricks. Her heaviness bends the blues and lights upon her back, and courses them through her diminutive scales made of teeth—her splendid, aerodynamic mithril, gray and cutting, excellent powers essential to her waterbending. Her back as a whole is beautiful and big—monolithic, in fact: a steady, massive, deep granite grey, expansive and reassuring, affirming in all of her power, might, strength and muscle that she is present. One could soar above her back for hours, and spy the endless details upon her frame—beyond the triviality of parasites, or the history of scars—the tiny, gorgeous dots and nicks and ticks and nips that altogether make her unique amongst all others. Her place in the world is irreplicable. No other is like her exactly—not in all of perfection. She is massively more complicated than the delicacies of the snow, the intricacies of the leaf, and yet she moves as one mighty frame, bearer of the innumerable wonders. Scattered in tiny detail across her tiny scales are the tiny lights of lazuli—specks and flecks of blue present only in life, and robbed upon death. She wears a coat of blue stars, unperceivable by many, but magic to the talented few. She bears these hidden marks with hidden brilliance—a coat of blue stars coursing through bluer seas. Brilliance belongs solely to her, and her alone; her back is none's.

Rocketing upward from her massive grey, an unquestionably commanding fin. Large, triangular—falcate, some say—essential to the archetype of 'Shark'. Her fin is truly, doubtlessly, in the minds of all who perceive her, a Shark's Fin. To many, it is the flag of fear. To some, it is the mark of adventure. To her, it is a mystery. But it most certainly commands attention, respect, wonder, awe, admiration, fascination—it is an essential element: what is a Shark without its classic fin? To alter the fin would make her 'A Different Shark' but to remove the fin would make her 'Not A Shark'. This is her breast and phallus—her voluptuous, erect, attentive essential to her physicality. She is more than her fin, of course; but her fin is her breast and phallus. It is the mark upon minds, the command upon cultures—this sharpness, thickness, bigness and tallness that penetrates into the otherworldly, exciting the surface waters with terrific titillation.

But beyond this wondrous, gorgeous, climatic apex of her body, another fin kneels upon her back. Past her primary's tip, across a smooth and ridgeless column of skin, armor and muscle, a gentler slope rises to less arousing heights. This fin, however, is not secondary. Its size and shape are markedly different, but not less. Its form is more balanced—a steady and firm ascent, eloquently departing her body at the careful summit—the curving demarcation between fore and curling aft, the scooping edge of traceless falls. Splashing light then ripples upon a secret cove between this fin and her mighty tail. A secret place—a safe space—before the strong ascent of that powerful, sweeping, phantom of the beast—that cape to her design! Yes! To this billowing blade, this key so gravely intertwined with her absolute power: how could a shark bite or track without the behemoth's beat of her billowing blades—her cleaving caudal cutlass? Invisible vortices ripped into the water by way of this powerful pulsation—that mighty sword of a tail!

Its make is perfectly balanced and refined, with subtle webs, equal weights, upward arcs and silky frames. Its culmination: a trademark notch, indicative of a special motion—her way of carrying her self through the water. Sharply arcing around the fork of this magnificent gown (and weapon), a Southern blade: white, pearly, pure, pointing with poised sincerity to the submerging sands below. This is her rudder, in part, but equal to her motion as the cartilaginous upper frame—the Northern cleave arched from the heavens, pointing backwards—away, both rudders and masters of movement pushing her with their created vortices past chaotic breaks in the water unto new realities parted ahead. The flowering blades of her tail spring from a strong caudal peduncle, smoothly gradient from the magnificent grey to a pearl unbesmirched—a cream of white.

Her flank, her chest, her belly—it is all a creamy white. Pale is a terrible description, for it means "lack of color". Far from the truth, all beneath her is the sincere affirmation of a splendid, innocent cream: she was aptly named *leucas*, we made a good decision with our application. But her white is pure and full of color—not devoid, like pale, but exuberant—a soft but solid white or cream, shielded from the surface, but seen only by those who gaze upwards towards heaven. Her white is smooth, but edged by a shadowy silhouette, when one still gazes upward. Crescent wings are perhaps her most conspicuous feature—the falcate pectorals, much different in nature than her falcate 1st dorsal. These are sweeping, broad, but strong—still commanding—and steady: they are stability, but presence, much like a falcon's wings (but the falcon must be very large).

If one could for a moment, divert their upward attentions to her most sacred of places, one could enjoy and cry at the beauty of her small angel-wings. These adorn the unspeakable: no English can enter such a place, but for now, her wings are her guardians, keeping language in check. Birth becomes her, and baby Bulls. Past angel wings, their first views of the world—the blue, the sand, the sky, the sea, the calm, the dancing, the tides, their pull—the light! Past these, guardians between mother and matter, the first dive of life! Their eyes illuminated into amber by the omnipotence of the planet—alight, alive! Aware. These pups are not in the Portrait, but they could be: that is most precious, powerful, pure, preposterous, and perplexing—Perfection! Potentiality is a possibility; actuality is what is perceived, and potentiality is what is imagined, and she, with her angel wings, is—like many mothers—the guardian of imagination. And witness to this, before the blade of her tail, is yet another fin: a shadowy foil to the smooth incline of the 2nd dorsal above. This is a hidden anchor, secret in plain sight. It is the 9th fin.

Features create a sense of detail, but sensations such as these barely begin to love such unfathomable beauty and wonder. Her powerful, solid, shapely lower jaw, white and smooth and unbending—is home to a complex culture of sharply polished pyramids, precise in their execution of what mortal sacrifices may come into that massive maw. How many have pierced so deeply that red fell upon their pearly crowns? How many dipped as arrowtips into blood? How many dispatched the dragon scales of Tarpon, Mullet and Mackerel—silver armor stripped away by unbreaking points aligned solely with the sun? These hard, immortal elements, long intact past death and demise; they will turn black after she is gone, but now, and forever entwined with her life, they are white and brilliantly incandescent wonders of the crescent world. But they are also peaks.

For beyond their heights, a chasm—as if between two far distant ranges in the dark—an unavoidable, sinking blackness that would no doubt consume the lives of all who fell victim…down…down into that bottomless doom. This is the maw—the vast, unbridgeable passage to the nether: the stomach, the intestines, and the spiral valve—all hellish princes of destruction, demise, and disincorporation. This is the fire in her belly. This is the fire that animates her muscle and surges her blood with rich, seeping, boiling crimson. Tendons strengthen and organs burst with the vitality of this cruel design. Broken limbs and faces and fins dissolve into the bile and acid of hellish lakes (all hidden and scattered throughout the seas). These are the pools of life: begat from death. This is the heart of Evil.

But so much Goodness it brings! For above her maw and upper jaw (fitted with even larger peaks and passes jagged with cruel intention) is a crown of neurons collected into a single, electric mind. This central feature is face to her being—her consciousness, for she does perceive. Her eyes feed her mind with many wonders, and the consciousness repays them with an aptly challenging blackness at the center of her amber iris—a black hole of a miniature galaxy, unequal in size but comparable in bewilderment. How can one read that mystery in her eyes? How can one look at the untraceable details of her amber iris and be so foolish as to say, "Aha! I have penetrated the impenetrable blackness at her pupil! I *understand*!" No one does. No person can. Her consciousness is impenetrable; only a fool would claim that apish palms could grasp the impossible.

Her eyes slip through man's fingers. Their light is not fully seen, for man's own eyes are flawed and imperfect. What's more, her senses—in all their complexity—are impossible to match and perceive as she perceives. Can one imitate her Ampualle of Lorenzini? No: it is impossible. No one can feel the electromagnetic surge throughout her body: she is uniquely gifted on that accord. No one can hear underwater as she can, or feel the Earth's polarity, or taste blood in the water, or see sunlight ripple in the spectrum of her vision. Imagine what scents plunge into her being! Day after day, a new stimulation, coursing through the coiled passages of her nares: could one but handle so many scents of the sea? Man can't even breathe underwater, let alone perceive with his limited sensuality. She is matchless. Her world is beyond comprehension. Forever.

To believe that we can fully *understand*—in the way I meant to try—is to see voices in the dark. We can't comprehend: it's impossible. But what is the value, then? What are the layers of *understanding*, sub-perfection? For clearly, there is a gradient between simple understanding and total *understanding*: a simple mind would call a Bull Shark 'big and dangerous' while a complex mind would add, 'present in Baghdad, high testosterone, central to Indonesian folklore, teeth date back to millions of years before our time'. There is value in the ascension of the human mind to the inner spheres of perceivable potential: it is far better to try to reach *understanding*—though it is impossible—than to satisfy one's self with basal knowledge. If *understanding* is akin to a star, then human potential is akin to a telescope: yes, we could satisfy our selves with our natural eyes and their limited perception—basal knowledge—but we have the potential to enhance our view and arrive much closer to the truth—to complex knowledge: the maximized attempt of harmonization with true *understanding*. This could be of any subject, of course, but to us, we apply ourselves to what we call *Carcharhinus leucas*.

Carcharhinus leucas is code for a being entirely within itself, and thus the ultimate descriptor of this being is *Carcharhinus leucas* (as far as human potential goes). *Carcharhinus leucas* is not the sum of its attributes—a complex mix of qualities such as its euryhaline nature and dangerous history—but instead, its own entity as best described by *Carcharhinus leucas*. In simplest terms: *Carcharhinus leucas* is *Carcharhinus leucas*.

And what is *Carcharhinus leucas* to you, oh beholder of the Portrait? What do *you* see? Your experience throughout our journey has resonated in a unique and immeasurable way: *your perspective* is unmatchable. Perhaps it was the Bull Shark's cannibalistic nature that haunts you the most? Or its mysterious history in the Miocene Mediterranean—in relation to the Messinian Salt Crisis, that catastrophic natural disaster? What has resonated with you specifically? What is *your* takeaway from this?

I do not know what is mine—or rather, English may not be useful enough for an accurate description. I have studied this species for two years, and I am beginning. Presence is something I feel...not excitement, nor enchantment, but *presence*. I am aware—violently aware—of this being, unique to me amongst all Sharks, for I have taken the time to perceive, to discover, to practice, to immerse myself—to feel the gravity of its presence, like the Earth feels the Moon. Scientifically, my relationship with this being goes unreciprocated...but spiritually, in line with The Ultimate or The Way, do I bear some *significance with* this animal? Am I destined to approach *Carcharhinus leucas*? Are you? You embraced this story; you read, you discovered, you joined—our triple existences are intertwined. We have mingled three separate threads anchored by one living truth: your perspective, my perspective, the collective of the scientific community, all huddled around the fire of the Bull Shark being—the *Carcharhinus leucas*.

There are no answers to questions beyond the empirical, and there perhaps shouldn't be—and certainly *wouldn't* be (at least in the universal sense) given our limited perspectives. But *Carcharhinus leucas* is a power unique among all others—no other *Carcharhinus leucas* exists in the entire cosmos. *Carcharhinus leucas* is *Carcharhinus leucas* forever: immortal, unchanging, an unbendable identity across time and space. Even in death— like *Carcocles megalodon*—the identity of *Carcharhinus leucas* is immortal and irreplaceable. This uniqueness is special in itself—it is a priceless assemblage of atoms and energy whose nature obliterates the coldness of such empirical description. What violent, graceful, bloody, quiet, tropical, massive, beautiful life it affirms! And this energy—this spirit of the universe, and more intimately, of the world—is by our estimations Near Threatened. We are pushing its existence—the priceless, fragile, vulnerable uniqueness that is the very same as your child's, but on a global scale—into the Perilous Realm: the expansive, accommodating cliffs adjacent to the bottomless abyss of Oblivion.

Do you understand Oblivion? Final, total, complete nonexistence; *Carcharhinus leucas*, upon entering Oblivion, will never return to this world. Extinction is Oblivion, but the term has been used too lightly, to frivolously, and too abusively by political lies: the real danger of Extinction escapes common grasp—*it is Oblivion*. The entire genetic line that created *Carcharhinus leucas* could never survive Oblivion. The present genetic lines could never re-create *Carcharhinus leucas*: that is impossible. Oblivion is all consuming: the Bull Shark may mirror its power in the form of predation, but such act is miniscule compared to Oblivion's total awe— its perfect nothingness that turns atoms inward into nonexistence. Oblivion will be the only thing to kill *Carcharhinus leucas*.

The Portrait! Look! Embrace this beautiful, gorgeous wonder, for it *exists now*! Beauty is of the present Universe! *Carcharhinus leucas* sits comfortably upon our tongues, and enriches our eyes with inescapable wonder. Pleasure pours throughout our being in witness to the wonder of this perfect imperfection—this unique magnificence. Blues dance, lights dance, tails pulse, eyes flash and roll, perceiving, hunting, searching, thinking—being in electrifying mystery, inescapable wonder—always *beyond our reaching hands.* So is one truth to the Bull Shark, and to all elements of this marvelous world…

The Bull Shark pulls flesh
To compound its heaviness
And push the waters

APPENDIX: CHART OF WORKS

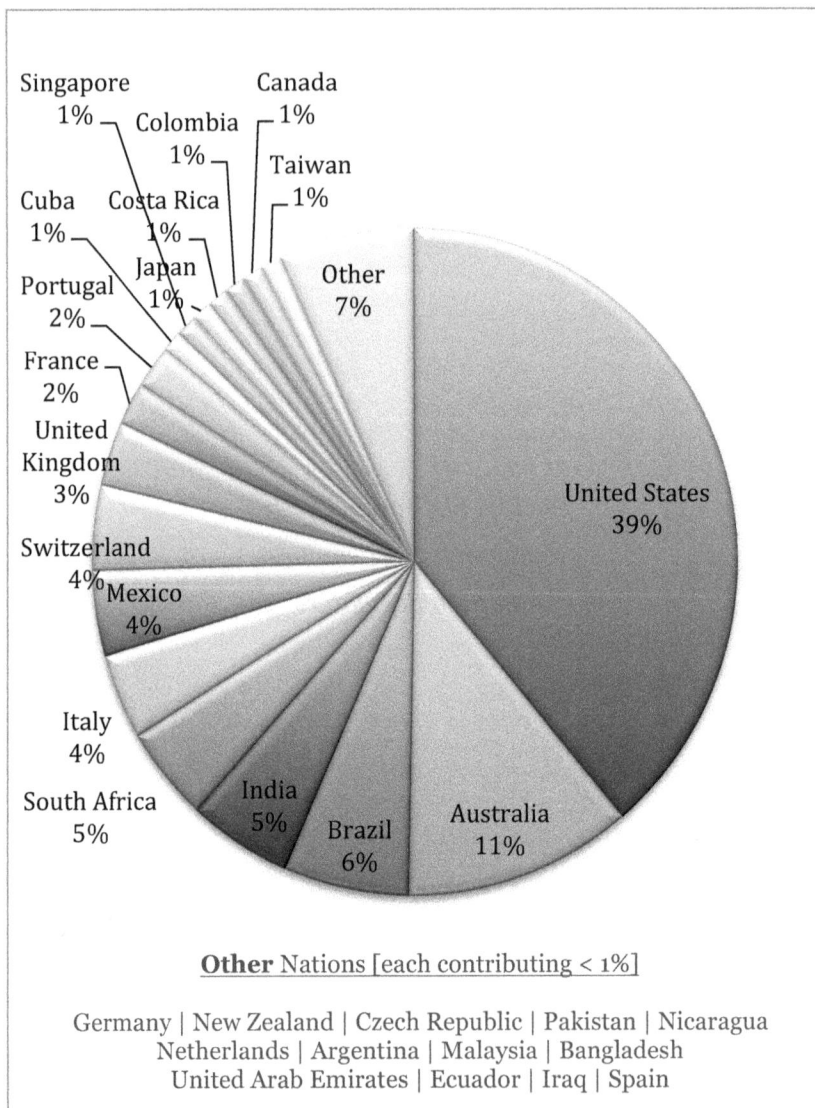

Singapore 1%
Colombia 1%
Canada 1%
Taiwan 1%
Cuba 1%
Costa Rica 1%
Japan 1%
Portugal 2%
France 2%
United Kingdom 3%
Switzerland 4%
Mexico 4%
Italy 4%
South Africa 5%
India 5%
Brazil 6%
Australia 11%
Other 7%
United States 39%

Other Nations [each contributing < 1%]

Germany | New Zealand | Czech Republic | Pakistan | Nicaragua
Netherlands | Argentina | Malaysia | Bangladesh
United Arab Emirates | Ecuador | Iraq | Spain

This chart depicts the % contribution of each nation to the total collection of empirical works (papers, articles, entries, guides and other scientific or journalistic publications) featured within the Bull Shark Compendium.

INDEX OF PRIMARY AUTHORS

LIST OF WORKS

W

Z

SUPPLEMENTARY SOURCES

1: Morgan, Alexia C., and George H. Burgess. "The Commercial Shark Fishery Observer Program: History, collection methodology and summary statistics 1994-2005(1)." Shark SEDAR Data Workshop Document. University of Florida, Nov. 2005. Web. 23 May 2016.

2: Noorden, Richard V. "The True Cost of Science Publishing." Editorial. *Nature* 495 (2013): 426-29. Web. 5 June 2016.

3: Swoger, Bonnie. "Why are journals so expensive?" *Scientific American* Sept. 2012.Web. 5 June 2016.

4: Gonzalez, Robbie. "Why does it cost $20,000 a year to subscribe to a science journal?." *Gizmodo*. Ed. Katie Drummond, Alex Dickinson, Rob Bricken, Jennifer Ouellette, and Michael Nunez. N.p., 21 Sept. 2011. Web. 5 June 2016. http://io9.gizmodo.com/5842304/why-does-it-cost-20000-a-year-to-subscribe-to-a-science-journal

5: Wallace, L. "Reactions of the Sharks Carcharhinus leucas (Müller & Henle) and Odontaspis taurus (Rafinesque) to Gill Net Barriers under Experimental Conditions." *The Oceanographic Research Institute Investigational Report* 30 (1972): 1-24. Web. 7 June 2016.

6: *The IUCN Red List of Threatened Species*. IUCN, n.d. Web. 16 July 2016. http://maps.iucnredlist.org/map.html?id=39372

7: Hoekstra JM, Molnar JL, Jennings M, Revenga C, Spalding MD, Boucher TM, Robertson JC, Heibel TJ, Ellison K (2010) The Atlas of Global Conservation: Changes, Challenges, and Opportunities to Make a Difference (ed. Molnar JL). Berkeley: University of California Press.

8: *Beqa Adventure Divers*. N.p., 2015. Web. 28 July 2016. <http://fijisharkdive.com/shark-info/our-big-fish-by-name/our-bulls/>.

9: Cliff G, Dudley SFJ (1991) Sharks caught in the protective gill nets off Natal, South Africa. 4. The bull shark *Carcharhinus leucas* Valenciennes. SA. J Mar Sci 1: 253–270.

10: Jackson AL, Parnell AC, Inger R, Bearhop S (2011) Comparing isotopic niche widths among and within communities: SIBER–Stable Isotope Bayesian Ellipses in R. J Anim Ecol 80: 595–602. doi:10.1111/j.1365-2656.2011.01806.x. PubMed: 21401589.

11: Lucifora, L. O., V. B. García, and B. Worm. 2011. Global diversity hotspots and conservation priorities for sharks. PLoS ONE 6:e19356.

12: *Food and Agriculture Organization of the United Nations* www.fao.org/fishery/species/2812/en. Accessed 1 Nov. 2016.

13: "Iraqi fisherman nets shark in Euphrates." *Middle East Online*, Middle East Online, Oct. 2007, www.middle-east-online.com/english/?id=22825. Accessed 7 Nov. 2016.

14: "Iraqi fisherman nets shark 160 miles from sea." *Reuters* Oct. 2007, www.reuters.com/article/us-iraq-shark-odd-idUSL3019334620071030. Accessed 9 Nov. 2016.

15: Tillett BJ, Meekan MG, Field IC, Thorburn DC, Ovenden JR. Evidence for reproductive philopatry in the bull shark *Carcharhinus leucas*. J Fish Biol. 2012;80: 2140–2158. doi: 10.1111/j.1095-8649.2012.03228.x. pmid:22551174

16: Eschmeyer WN, Fong JD (2015) Species by family/subfamily. Department of Ichthyology, California Academy of Sciences. www.calacademy.org/research.calacademy.org/research/ichthyology/catalog/SpeiesByFamily.asp

17: Pirog A, Blaison A, Jaquemet S, Soria M, Magalon H. 2015. Isolation and characterization of 20 microsatellite markers from Carcharhinus leucas (bull shark) and cross-amplification in Galeocerdo cuvier (tiger shark), Carcharhinus obscurus (dusky shark) and Carcharhinus plumbeus (sandbar shark). Conservation Genetics Resources 7:121–124 DOI 10.1007/s12686-014-0308-3.

ACKNOWLEDGMENTS

COVER PHOTO

Bull Shark 069 by Andy Murch, Elasmodiver

Many thanks to Andy Murch and the Elasmodiver team for their stunning photograph of *Carcharhinus leucas*. Elasmodiver photographs by Andy Murch include some of the world's best profiles of sharks and rays, and are specifically chosen for their exquisite balance of diagnostic scope and aesthetic intimacy. For more information on Elasmodiver's incredible photographic collection, please visit: **www.elasmodiver.com**

PAPERS

While all of the papers featured within *The Bull Shark Compendium* can be found online via an incalculable variety of disparate sources, there is one website in particular that has taken the unparalleled initiative to compile the addresses, links and citations of each into a single convenient (and critically important) database: many thanks are thus due to Shark-References.com by Jürgen Pollerspöck & Nicolas Straube (in cooperation with the Bavarian State Collection of Zoology in Munich). Shark-References must be one of the 21st Century's greatest innovations for shark research and study, and should be of biblical importance to shark scientists, conservationists and enthusiasts alike. For more information on this exhaustive and powerful resource, please visit: **www.shark-references.com**

ABOUT THE AUTHOR

"Peter Benchley, author of the book that inspired the movie Jaws, was well aware that sharks have a public relations problem. He spent many years combatting the misconceptions that have made shark conservation such a tricky issue—namely, the idea that all sharks are giant, ruthless killing machines intent on attacking every human they encounter....The late author would no doubt have approved of a new series of books, created by the "hybridist author" Dr Jaws, which disabuses readers of these sorts of inaccuracies and celebrate sharks for what they are— beautiful and fascinating animals that are an important part of their ecosystem and key components of many cultures around the globe."

~ Dr. Caitlin Kight, Editor, Current Conservation Magazine

(Exeter, Cornwall, UK)

Dr. Jaws is a brand of shark books and media that hybridizes art, science, philosophy and ocean lore into comprehensive presentations of U.S. Mid-Atlantic shark species. Created in 2012 by Zachary Webb Nicholls (College of William & Mary, B.S. Biology, '14) , Dr. Jaws has since expanded into a series of seven books, a podcast (*Sharks & Coffee*), an educational lecture circuit (*Sharks of the Chesapeake*) and an extensive collection of sharky tweets, posts and videos on Twitter, Facebook and YouTube. Dr. Jaws is recognized as an 'informal educator' by both the National Marine Education Association (NMEA) and its Mid-Atlantic chapter (MAMEA), and has been internationally incorporated into schools, aquariums and expos in the United Kingdom, South Africa and Hong Kong.

For more information on Dr. Jaws, please visit:

www.drjaws.net

Photograph by Ian P. Herbst